IUFRO Research Series

The International Union of Forestry Research Organizations (IUFRO), with its 14,000 scientists from 700 member institutions in 100 countries, is organized into nearly 300 research units that annually hold approximately 60 conferences, workshops and other meetings. The individual papers, proceedings and other material arising from these units and meetings are often published but in a wide array of different journals and other publications. The object of the IUFRO Research Series is to offer a single, uniform outlet for high quality publications arising from major IUFRO meetings and other products of IUFRO's research units.

The editing, publishing and dissemination experience of CABI *Publishing* and the huge spread of scientific endeavours of IUFRO combine here to make information widely available that is of value to policy makers, resource managers, peer scientists and educators. The Executive Board of IUFRO forms the Editorial Advisory Board for the series and provides the monitoring and uniformity that such a high quality series requires in addition to the editorial work of the conference organizers.

While adding a new body of information to the plethora currently dealing with forestry and related resources, this series seeks to provide a single, uniform forum and style that all forest scientists will turn to first as an outlet for their conference material and other products, and that the users of information will also see as a reliable and reputable source.

Although the official languages of IUFRO include English, French, German and Spanish, the majority of modern scientific papers are published in English. In this series, all books will be published in English as the main language, allowing papers occasionally to be in other languages. Guidelines for submitting and publishing material in this series are available from the Publisher, Books and Reference Works, CABI Publishing, CAB International, Wallingford, Oxon OX10 8DE, UK, and the IUFRO Secretariat, c/o Federal Forest Research Centre, Seckendorff-Gudent-Weg 8, A-1131, Vienna, Austria.

Titles Available:

1. *Forest Dynamics in Heavily Polluted Regions*
   Edited by J.L. Innes and J. Oleksyn

2. *Forest History: International Studies on Socio-economic and Forest Ecosystem Change*
   Edited by M. Agnoletti and S. Anderson

3. *Methods and Approaches in Forest History*
   Edited by M. Agnoletti and S. Anderson

4. *Air Pollution and the Forests of Developing and Rapidly Industrializing Countries*
   Edited by J.L. Innes and A.H. Haron

5. *Forests in Sustainable Mountain Development*
   Edited by M. Price and N. Butt

Air Pollution and the Forests of Developing and Rapidly Industrializing Regions

Report No. 4 of the IUFRO Task Force on Environmental Change

Edited by

J.L. Innes
Department of Forest Resources Management, University of British Columbia, Vancouver, Canada

and

A.H. Haron
Forestry Department Peninsular Malaysia, Kuala Lumpur, Malaysia

CABI *Publishing*
in association with
The International Union of Forestry Research Organizations (IUFRO)

CABI *Publishing* is a division of CAB *International*

CABI Publishing
CAB International
Wallingford
Oxon OX10 8DE
UK

Tel: +44 (0)1491 832111
Fax: +44 (0)1491 833508
Email: cabi@cabi.org
Web site: http://www.cabi.org

CABI Publishing
10E 40th Street
Suite 3203
New York, NY 10016
USA

Tel: +1 212 481 7018
Fax: +1 212 686 7993
Email: cabi-nao@cabi.org

A catalogue record for this book is available from the British Library, London, UK.

Library of Congress Cataloging-in-Publication Data
Pollution and the forests of developing and rapidly industrializing regions/edited by J.L. Innes and A.H. Haron.
p. cm. -- (IUFRO research series; no. 5)
Includes bibliographical references.
ISBN 0-85199-481-4 (alk. paper)
1. Forests and forestry--Environmental aspects--Developing countries. 2. Forest ecology--Developing countries. 3. Air--Pollution--Environmental aspects--Developing countries. I. Innes, John L. II. Haron, A.H. (Abu Hassan) III. Series.

SD387.E58 P65 2000
333.75′14′091724--dc21 00-023662

Published in Association with:

The International Union of Forestry Research Organizations (IUFRO)
c/o Federal Forest Research Centre
Seckendorff-Gudent-Weg 8
A-1131 Vienna
Austria

ISBN 0 85199 481 4

Typeset by AMA DataSet Ltd, UK
Printed and bound in the UK by the University Press, Cambridge

Contents

Contributors

N. Abdul Rahim, Forest Research Institute Malaysia, Kepong, 52109 Kuala Lumpur, Malaysia

M. Agrawal, Department of Botany, Banaras Hindu University, Varanasi-221 005, India

S.B. Agrawal, Department of Biological Sciences, Allahabad Agricultural Institute, Allahabad-211 007, India

R. Atmawidjaja, Universitas Pakuan, JL Pakuan, PO Box 353, Bogor, Indonesia

L.I. de Bauer, Instituto de Recursos Naturales, CP 56230 Montecillo, Mex. Mexico

N. Chacón, Centro de Ecología, Instituto Venezolano de Investigaciones Científicas, Caracas, Venezuela

T. Dahlan, Regional Centre for Forest Management (RCFM), B-11-8, 10 and 11 Block B, 11th Floor, Megan Phileo Avenue, 12 Jalan Yap Kwan Seng, 50450 Kuala Lumpur, Malaysia

T.F. Fameso, Forestry Management, Evaluation and Co-ordinating Unit (FORMECU), PMB 226, Wusse, Zone 1, Abuja FCT, Nigeria

A.H. Haron, Forestry Department Peninsular Malaysia, Jalan Sultan Salahuddin, 50660 Kuala Lumpur, Malaysia

T. Hernandez-Tejeda, Instituto Nacional de Investigaciones Forestales, Agricolas y Pecuarias, SAGAR, Av. Progreso No. 5, 04110 México DF, Mexico

R. Herrera, Departamento de Química, Universidad Simón Bolívar, Caracas, Venezuela

W.K. Hicks, Stockholm Environment Institute at York, University of York, PO Box 373, York YO10 5YW, UK

J.L. Innes, Department of Forest Resources Management, University of British Columbia, Forest Sciences Centre, 2045, 2424 Main Mall, Vancouver, British Columbia V6T 1Z4, Canada

M. Iyngararasan, UNEP Environment Assessment Programme for Asia and Pacific, AIT, PO Box 4, Klongluang, Pathumthani 12120, Thailand

Y.-K. Kim, Air Pollution Laboratory, Forestry Research Institute, 207 Cheongryangri-dong, Dongdaemun-ku, Seoul 130-012, Korea

J.C.I. Kuylenstierna, Stockholm Environment Institute at York, University of York, PO Box 373, York YO10 5YW, UK

V. Mathur, Stockholm Environment Institute at York, University of York, PO Box 373, York YO10 5YW, UK

S. Mazzucchelli, Instituto Internacional de Medio Ambiente y Desarrollo (IIED), America Latina, Av. Gral. Paz 1180, Buenos Aires, Argentina

D.I. Odintsov, Federal Forest Service of Russia, Pyatnitskaya str. 59/19, Moscow 113184, Russia

K.A. Olbrich, PO Box 12716, Nelspruit 1200, South Africa

D. Mohd Puat, Regional Centre for Forest Management (RCFM), B-11-8, 10 and 11 Block B, 11th Floor, Megan Phileo Avenue, 12 Jalan Yap Kwan Seng, 50450 Kuala Lumpur, Malaysia

D.F. Richards, Universitas Pakuan, JL Pakuan, PO Box 353, Bogor, Indonesia

H. Shang, Research Institute of Forest Ecology and Environmental Science, Chinese Academy of Forestry, Wan Shou Shan, Beijing 100091, China

S. Shrestha, UNEP Environment Assessment Programme for Asia and Pacific, AIT, PO Box 4, Klongluang, Pathumthani 12120, Thailand

J.M. Skelly, Department of Plant Pathology, The Pennsylvania State University, University Park, PA 16803, USA

A.M. van Tienhoven, CSIR Environmentek, PO Box 395, Pretoria 0001, South Africa

D. Xu, Research Institute of Forest Ecology and Environmental Science, Chinese Academy of Forestry, Wan Shou Shan, Beijing 100091, China

Foreword

The quest for sustainable development has gained increasing momentum since the 1992 UN Conference on Environment and Development. For many countries, the development of their economies has been accompanied by environmental and ecological costs. The pattern of environmental damage seen in Western Europe and North America as countries in these regions industrialized is being repeated elsewhere as countries strive to increase the standard of living of their inhabitants.

Environmental pollution can be one of the ecological costs of development. Such pollution comes not only from industry but also from forest fires, as so graphically illustrated by the problems in South-east Asia during the late 1990s. As discussed in some of the chapters in this book, pollution can also have social and economic costs, with major impacts on human health. Pollution control is often driven first by concerns about the effects on the population. Once human health is safeguarded, concern may then turn to the environment.

Forests represent an important feature of many rapidly industrializing countries. While the primary concern has been for the rates of forest loss, changes in forests induced by air pollution are likely to become increasingly important if air pollution control policies are not implemented. There is a major gap in our understanding of air pollution impacts on forests, and this book will help to identify some of the most important. IUFRO scientists can make an extremely valuable contribution to improving the environment, and I would encourage the global collaboration of scientists concerned with these environmental problems.

Professor Jeff Burley
President, International Union of Forest Research Organizations

Air Pollution and Forestry in Rapidly Industrializing Countries: an Introduction

1

J.L. Innes[1] and A.H. Haron[2]

[1]Department of Forest Resources Management, University of British Columbia, Vancouver, British Columbia, Canada; [2]Forestry Department Peninsular Malaysia, Kuala Lumpur, Malaysia

The forests of developing and rapidly industrializing regions and countries are under many pressures. The historical legacies of individual countries mean that the relative contributions to forest degradation of logging, grazing and firewood collection differ. Air pollution represents an additional pressure, but very little is known about its impacts in most regions outside Europe and North America. Current concerns about air pollution have focused on its impacts on the health of urban populations in centres such as Mexico City. In some heavily polluted regions, such as around Chongqing in China, there have been intensive studies of the ecological impacts of pollution. However, for many countries, reducing the impacts of air pollution on forests has a much lower priority than more immediate issues such as poverty alleviation and the provision of food. Despite this, information is accumulating that suggests that the impacts on forests may be substantial, with important implications for the future health of the forests and the maintenance of biodiversity in protected areas.

1 Introduction

Sustainable development is an issue that has received increasing global attention, largely as a result of the activities of the UN Commission on Sustainable Development. For countries throughout the world, the development of their economies remains an important priority. This is especially true of developing and rapidly industrializing countries, where the creation of wealth is a key factor in the alleviation of many social, economic and environmental problems. However, such development often comes at a cost: environmental pollution. The most immediate problems associated with environmental pollution are the effects on human health. These have been at their most obvious following media exposure of accidents, such as the 1984 accident at the Union Carbide plant in Bhopal, India. In developed countries, the effects of pollution on health have been recognized for many years, resulting in strict air pollution control legislation. After resolving the problems of human health, attention turned to the effects on crops of economic importance. Finally, attention has focused on the effects of pollution on vegetation of no economic significance and on issues such as visibility in recreational areas.

Many developing countries are just starting out on this path. Priority is given to reducing health hazards and the protection of vegetation is seen as a luxury that at present is fairly low in the list of priorities. Some industrial developments involve the use of hazardous technology, and the full risks associated with the technology may be inadequately understood or taken into account (James, 1991). However, with increasing globalization, people in developed countries are attempting to place their values on what is happening in developing countries. This is creating conflict, as illustrated by the ongoing debates about who should pay for reductions in carbon dioxide emissions in developing countries. Some interpret attempts to restrict carbon dioxide emissions as a means by which developed countries can hinder the development of the economies of the rest of the world. However, the priority for many countries remains the creation of wealth while minimizing undesirable effects such as environmental pollution.

In examining the relationships between forests and air pollution in developing and rapidly industrializing countries, a number of different themes are evident. The greatest concerns relate to the links with human health. Forest fires and other forms of biomass burning (especially the burning of woodfuels in domestic situations) represent a significant health hazard in some areas. The large-scale fires that have occurred in Indonesia, Brazil and elsewhere in recent years are identified in several chapters (Chapters 2, 5, 8 and 11) in this book as being important, and considerable emphasis has been placed on these. The burning of woodfuels is given less attention, although a short review of the problem has been included (Chapter 10) to ensure that its overall importance is not forgotten (see for example Florig, 1997).

2 Pollution associated with point sources

One of the best-documented cases of air pollution impacts in the regions under consideration comes from the Chongqing area of Sichuan Province in south-west China. Chongqing is a city of more than 2.5 million people and is the biggest industrial and commercial centre in south-west China (Zheng *et al.*, 1998). This area is badly polluted as a result of the combustion of coal with a high sulphur content; the local coal contains 3–5% sulphur (Zhang *et al.*, 1996). An annual mean concentration of SO_2 of 220 μg m^{-3} (77 ppbv) has been reported, as have annual mean concentrations of particulate sulphate (SO_4^{2-}) of 32 μg m^{-3} (Shen *et al.*, 1995). Bulk deposition to Masson pine (*Pinus massoniana*) forests has been estimated to be 932.1 kg S ha^{-1} in throughfall and 46.6 kg S ha^{-1} in stemflow (Shen *et al.*, 1995). These rates are amongst the highest recorded anywhere in the world. The acidic deposition is also accompanied by mercury deposition, and elevated mercury levels have been identified in soils, crops and milk (Mou and Qing, 1995). Fluoride also appears to be involved, with elevated foliar contents of fluoride having been identified in some studies (Yu *et al.*, 1988, 1990a,b).

Numerous studies around Chongqing have indicated that air pollution has major effects on the soils and trees there (e.g. Liu *et al.*, 1988a,b; Zhao and Seip, 1991; Liang *et al.*, 1994). Symptoms on Masson pine include tip necrosis on needles, yellow flecking, reduced foliar biomass, reduced needle length, premature abscission, branch dieback, reduced root growth and reduced radial growth (Liu *et al.*, 1988b; Yu *et al.*, 1990a). Symptoms on broadleaved trees, including black locust (*Robinia pseudoacacia*), Daimyo oak (*Quercus dentata*), swamp mahogany (*Eucalyptus robusta*) and *Eucalyptus variegata*[1], include necrotic lesions, although broadleaved species appear to be less sensitive than conifers (Yu *et al.*, 1990b). The deposition under Masson pine is higher than under mixed coniferous forest or under forests of camphor trees (*Cinnamomum camphora*), and the vegetation also influences the extent of aluminium dissolution (Okazaki *et al.*, 1998). However, as the measurements taken to reach these conclusions were made at different locations (Mt Zhenwu for Masson pine, Mt Jinyun for the mixed coniferous forest and Laojundong for the camphor trees), site interactions may have played a role. The acidic deposition associated with the pollution appears to have more impact on Masson pine than on camphor trees, causing a reduction in increment in the pine (Wu, 1998).

Sulphur dioxide is a significant source of vegetation injury in a number of other areas. For example, forest damage attributable to SO_2 has been identified in the Ulsan and Onsan areas (C.K. Lee *et al.*, 1998), Kunsan (Lee *et al.*, 1995) and Kwangyang Bay (Nagaya *et al.*, 1998) of Korea. Damage has been identified on a number of tree and shrub species, including Manchurian fir (*Abies holophylla*), persimmon (*Diospyros kaki*), Norway spruce (*Picea abies*) and Japanese red pine (*Pinus densiflora*) (Song *et al.*, 1996; Nagaya *et al.*, 1998), and on crops such as *Zea mays*. Sulphur dioxide is a major issue in parts of Siberia and eastern Russia (Kharuk, 1999; see also Chapter 6), where large-scale smelters are important sources of SO_2 and heavy metals.

1 The taxonomic status of *Eucalyptus variegata* is uncertain. It is not accepted as a valid species, and is either part of *E. citriodora* or a synonym for *Corymbia variegata* (F. Muell.) K.D. Hill & L.A.S. Johnson.

Fluoride is also an important problem in some areas, despite the availability of technology to reduce fluoride emissions to acceptable levels. For example, fluoride is one of several phytotoxic pollutants affecting the Atlantic rainforest around the Cubatão industrial complex near São Paulo, Brazil (Klumpp *et al.*, 1996a,b, 1997). In this case, the fluorides are emitted by fertilizer plants. Damage to a number of tree species has been documented, including *Tibouchina pulchra*, *Miconia pyrifolia* and *Cecropia glazioui*. Fertilizer factories are also important sources of pollution elsewhere. For example, a guide to the relative sensitivities of plants near fertilizer factories has been developed for the Philippines (Subido *et al.*, 1969). Similarly, damage caused by fluoride emitted from a fertilizer plant in the Sfax region of Tunisia has been documented (Ben-Abdallah and Boukhris, 1990). Symptoms of fluoride injury have also been described on western coastal wattle (*Acacia cyclops*), golden wreath wattle (*A. saligna*), sugar gum (*Eucalyptus cladocalyx*) and tuart (*Eucalyptus gomphocephala*) growing near a fertilizer plant at Cape Town, South Africa (Botha *et al.*, 1989). The two *Eucalyptus* species were more sensitive than the *Acacia* species.

Other important point sources include cement factories (Panes and Zamora, 1991; Iqbal and Iqbal, 1995) and lime kilns (Banerjee *et al.*, 1999), coal mines (Verma *et al.*, 1992), oil production areas (Morales *et al.*, 1995) and petroleum processing facilities (Salgare and Trisa, 1997), chemical plants (Suckcharoen, 1980; Vijayan and Bedi, 1989), and steel and other metallurgical plants (Nyangababo and Salmeen, 1987; Muaka *et al.*, 1989; Kharuk, 1999).

One specific symptom associated with point sources is black tip of mango (*Mangifera indica*), which occurs in the neighbourhood of brick kilns in India. The symptom is believed to be induced by a combination of sulphur dioxide, ethylene and carbon monoxide (Rajput *et al.*, 1971; Ram, 1989). It can be alleviated by spraying fruits with NaOH (0.6–0.8%) or NaCO (0.6%) solutions, or by planting tolerant cultivars such as 'Langra' (Ram, 1989). However, as in all such cases, the best solution would be to reduce the levels of pollution.

Natural sources of pollution may be important in some areas. Volcanoes are a significant source of sulphur and other pollutants and large areas may be damaged by ash falls (Schmitt and Whittaker, 1998). Very few data are available, but the 12 active volcanoes in Japan at the end of the 1980s were estimated to emit 1.5 Mt SO_2 annually (Ichikawa *et al.*, 1997). Biogenic sources of sulphur may also be important, as in the Sinamaica Lagoon–Lake Maracaibo area of Venezuela (Morales *et al.*, 1995).

3 Pollution associated with diffuse sources

The biggest pollution problem associated with diffuse sources is ozone. Ozone (O_3) is a secondary pollutant arising from interactions between nitrogen oxides (comprising nitric oxide and nitrogen dioxide), reactive hydrocarbons and sunlight. In sunlight, NO_2 decomposes in air to form NO, and the NO reacts with the hydrocarbons to form O_3. In many cities, large amounts of nitrogen oxides are emitted from traffic: the problem is particularly apparent in areas where two-stroke engines are used extensively. Ozone is also formed from the pollutants released by biomass burning, and ozone episodes in areas such as south-west Africa, formerly considered to be pristine, are now being documented. Ozone is a phytotoxic gas that also has adverse effects on human health. It has been recognized as phytotoxic for over 40 years, and one of the best-known bioindicators is tobacco (*Nicotiana tabacum*) (particularly the Bel W3 cultivar). It is therefore interesting to note that tobacco plants in Cuba have been recorded with the typical symptoms of ozone injury (originally known as 'weather fleck') and that the symptoms have been reproduced experimentally during ozone treatments (Quintero and Ramirez, 1990).

The extent to which ozone represents a problem in countries such as China is very uncertain. For example, at Chongqing, peak hourly mean ozone concentrations of 93 ppb have been recorded (Zheng *et al.*, 1998), indicating that a problem exists. While this is insufficient to cause serious impacts, a continued increase in the severity of ozone impacts could have marked effects. Elsewhere, ozone is already likely to be having effects. For example, ozone concentrations in the eastern Mpumalanga region of South Africa (an area with large areas of commercial forest plantations) are sufficiently high to cause problems (Carlson, 1997), although effects have not yet been documented (see Chapter 5). In Pakistan, high 6 h mean O_3 concentrations have been reported and these are believed to be having an impact on the yield of wheat and rice (Maggs *et al.*, 1995).

Urban pollution can be identified as an increasingly serious problem. It is always difficult to say whether this represents a point or diffuse source as it depends on the scale of the observation. The pollution is derived from multiple sources, including cars, industry and power generation, but the sources are concentrated in a relatively discrete area. Cities in rapidly industrializing areas with significant problems include Seoul (Korea) (Hyun, 1997), Santiago de Chile (Rutllant and Garreaud, 1995), São Paulo (Brazil), Caracas (Venezuela) (Gordon *et al.*, 1995), Kathmandu (Nepal) (Dhital, 1994; Sharma, 1996), Karachi (Pakistan) (Beg and Shams, 1989), Chongqing (China), Calcutta (India) and Bangkok (Thailand). Urban pollution takes many forms, although attention in rapidly industrializing countries has focused on ozone, fluorides and heavy metals. Heavy metals, particularly lead, are of concern because of their well-documented effects on human health. Many studies have been conducted (e.g. in Benin City, Nigeria: Ademoroti, 1986; Jos, Nigeria: Fatoki, 1987; Sharkiya, Egypt: El-Desouky *et al.*, 1998), all of which indicate that heavy metal pollution can be a problem.

Many of the problems in cities are exacerbated by the local topography. The classic example is the Mexico City Basin, but other cities are equally susceptible. In Santiago de Chile, for example, pollution episodes are mainly associated with the leading edges of coastal lows that lower the base of the semi-permanent temperature inversion, thereby reducing the diurnal growth of the surface mixed layer (Rutllant and Garreaud, 1995). The problems of Mexico City are examined in detail in Chapter 3, where the socio-economic and historical issues that led to the current situation are discussed. Air pollution from road traffic appears to be a problem in a number of areas, although care is needed to ensure that any symptoms are correctly ascribed to air pollution. A number of other stresses can cause damage to roadside trees, including winter salting, drought and root restriction and/or root removal. Khan *et al.* (1995) have described symptoms on *Alstonia scholaris*, *Heterophragma adenophyllum*, *Mimusops elengi*, *Ficus religiosa*, *F. benghalensis* and *F. infectoria* growing in roadside situations along the Lahore Mall (Pakistan). The symptoms included senescence, thinning of the canopy, changes in phenology, changes in leaf size, leaf injury and dust deposition. Similarly, in Karachi (Pakistan), impacts on species such as *Guaiacum officinale*, *Peltophorum pterocarpum* and *Albizia lebbeck* have been noted (Iqbal *et al.*, 1997).

4 Forest fires as a source of pollution

Forest and savannah fires, both natural and of anthropogenic origin, have received a great deal of attention in recent years. They are important both as a health hazard in the region of the fire and globally as a source of greenhouse gases and as a source of precursors for tropospheric ozone formation. The importance of forest fires as a source of pollution was brought to the world's attention by the Indonesian fires of 1997, 1998 and 1999. Extensive media coverage emphasized how cities in countries such as Malaysia and Singapore were being impacted. Pollutant concentrations well above those considered to be hazardous have been recorded in cities such as Kuala Lumpur during pollution episodes (Fang *et al.*, 1999).

Measurements taken during the 1998 fires in Indonesia revealed that air quality standards in a number of cities were violated. For example, in Brunei Darussalam, the World Health Organization (WHO) guideline of 70 $\mu g\ m^{-3}$ for PM_{10} (suspended particulate matter with a diameter of 10 μm or less) concentrations (24 h average) was exceeded on 54 days between 1 February and 30 April 1998 (Radojevic and Hassan, 1999). Other pollutants (including sulphur dioxide, nitrogen dioxide and ozone) did not exceed WHO guidelines, although the 8 h guideline for carbon monoxide was exceeded on 7 days. The daily average PM_{10} concentrations were generally below 450 $\mu g\ m^{-3}$, though concentrations in excess of 600 $\mu g\ m^{-3}$ and reaching up to 1 $mg\ m^{-3}$, persisting for several hours, occurred on a number of occasions (Radojevic and Hassan, 1999).

Tropical fires are also responsible for the acidification of precipitation (Andreae, 1991; Lacaux *et al.*, 1991). Organic acids, particularly formic and acetic acids, and nitric acid are primarily responsible for the acidity, which has been reported from Brazil, Venezuela and Africa. The acetic acid may be released directly during burning, and is also formed, along with formic acid, by photochemical reactions in the smoke plume (Helas *et al.*, 1991). Nitric acid is also formed photochemically, being derived from nitrogen oxides emitted in the fire. Andreae (1991) argues that these acids can reduce precipitation below pH 4.0.

Large-scale tropical fires have been identified for some time as having major atmospheric impacts. The most important gases released include carbon dioxide, carbon monoxide, methane and other hydrocarbons, nitric oxide and sulphur dioxide (Andreae, 1991). For example, in Brazil, the fires have been associated with regional increases in tropospheric carbon monoxide and ozone concentrations (Kirchhoff, 1996; Kirchhoff and Pavão, 1997). Ozone peaks of 138 ppbv in Hong Kong have been associated with air masses that have passed over continental South-east Asia where biomass burning has been occurring (Liu *et al.*, 1999).

A huge amount of information has been gathered about interactions between forest fires and climate thanks to a number of large-scale experiments. There have been several experiments conducted under the auspices of the International Global Atmospheric Chemistry (IGAC) project of the International Geosphere–Biosphere Program (IGBP). These include the Biomass Burning Experiment: Impact on the Atmosphere and Biosphere (BIBEX) that looked at grassland fires in southern Africa and Brazil, tropical forests in Brazil and the boreal forests of Russia. As part of BIBEX, the Southern Tropical Atlantic Regional Experiment (STARE) was conducted, a programme that included smaller (but still very large) experiments such as the Southern Africa Fire–Atmosphere Research Initiative 1992 (SAFARI–92) and Transport and Atmospheric Chemistry near the Equator–Atlantic (TRACE-A). An account of these experiments is provided in Levine (1996) and a recent account of biomass burning–climate interactions is provided by Innes *et al.* (1999).

5 Indoor wood-burning as a source of pollution

Large numbers of people depend on wood as their primary source of energy: in some countries, woodfuels meet more than 90% of domestic energy needs. Foresters have placed considerable emphasis on developing woodfuels as a renewable source of energy that has the potential to resolve some of the energy problems present in developing countries. However, wood smoke is a form of pollution, and when wood is burnt indoors, potentially harmful concentrations of pollutants can accumulate. Studies in a number of countries (e.g. Ellegard and Egneus, 1992) have indicated that exposure to smoke from cooking fires can have significant health effects. These issues are examined further in Chapter 10.

6 Changes in vegetation induced by pollution

There is limited evidence for widespread changes in vegetation associated with regional-scale pollution. In Korea, genetic differences between sensitive and tolerant Sargent cherry (*Prunus sargentii*) have been identified (S.W. Lee *et al.*, 1998), but this work has not been extended to the identification of changes in population genetics. Very little genetic work has been done elsewhere: even in the polluted regions of developed countries, only a limited amount of field-based work has been done on the impacts of pollution on population genetics.

In many cases, it is difficult to determine the community-level effects on vegetation as so little information is available on the forests themselves. An exception is provided by the detailed work that has been done around some pollution sources in India (see Chapter 9). As a result of the lack of information about forest ecology, studies are often limited to visible symptoms, changes in foliar chemistry and nutrition and changes in growth characteristics. An example is provided by the studies on the impacts of pollution on the Atlantic rainforest in the vicinity of São Paulo, Brazil. Here, despite the evidence of forest decline, work has been limited to some of the more obvious vegetation characteristics (Klumpp *et al.*, 1998). In some cases, such as with the tolerance to heavy metals of *Combretum psidioides*, detailed investigations have been made (Ernst, 1985). However, this is very much the exception, and the tolerance of most species to specific pollutants is very poorly understood.

There has been quite a lot of research undertaken using *in situ* bio-indicators. For example, Gordon *et al.* (1995) looked at metal contents in the lichen *Parmotrema madagascariaceum* growing on the bark of *Clusia multiflora* in cloud forests adjacent to the Caracas Valley (Venezuela). They identified elevated lead and zinc contents, indicating that the forests were exposed to these pollutants. Carreras *et al.* (1995) also looked at pollutant concentrations in lichens, working with transplanted *Usnea* spp. in Cordoba (Argentina). Similar results were obtained by Gonzalez and Pignata (1997), working with the transplanted *Punctelia subredecta* in Cordoba. In all cases, there was evidence of impacts, but the applicability of the results to other vegetation forms is uncertain. Roadside trees (Chinese privet – *Ligustrum lucidum*) in Cordoba have been examined (Carreras *et al.*, 1996; Canas *et al.* 1997), but these introduced, generally pollution-tolerant trees may not be representative of natural vegetation in the area. In Nigeria, the moss *Polytrichum juniperinum* and bark from the tree *Azadirachta indica* have been used as indicators of heavy metal pollution, although the biological impacts of the metals were not studied (Kakulu, 1993). In the Jharia Coalfield area of India, the moss *Funaria hygrometrica* has been used as a bioindicator, and here plant density rather than chemical content has been used. Pollution was found to cause significant declines in the occurrence of the species (Alok and Tewary, 1997).

There are many potential bioindicators that could be used for pollution studies. In many countries, the presence of the epiphytic bromeliads *Tillandsia* spp. presents an opportunity to monitor pollutants. For example, *T. caput-medusae* has been used successfully as a bioindicator in San José City (Costa Rica) (Brighigna *et al.*, 1997). Epiphytes potentially represent one of the best possible bioindicators and would be worth studying further.

7 Regional and transboundary pollution

Several authors in this book emphasize that pollution problems in rapidly industrializing countries are becoming increasingly regional and transboundary in nature. This has a number of ramifications. Documented examples include from the Korean peninsula to east China (Park and Cho, 1998) and from China to Korea and Japan (Wakamatsu *et al.*, 1996; Ichikawa *et al.*, 1997). In fact, pollution plumes from north-east Asia extend across the Pacific, influencing air quality as far away as British Columbia.

Biomass burning creates large-scale regional pollution problems. This is well illustrated by the pollution from the fires in Indonesia, but biomass burning in Africa (e.g. Douguedroit and Bart, 1997) and South America also results in transboundary pollution problems. For example, plumes from biomass burning in South America can extend as far as the west coast of Africa, where elevated concentrations of tropospheric ozone have been reported. Biomass burning actually appears to have global impacts, significantly affecting the amounts of some gases and aerosols in the earth's atmosphere (Granier *et al.*, 1997).

The presence of transboundary air pollution is important and has a number of implications. Both regional and transboundary pollution can hinder conservation efforts, if natural reserves are adversely impacted. For example, the Biological Reserve of Paranapiacaba in the Atlantic rainforest of Brazil is impacted by acidic deposition derived from emissions from the Cubatão industrial complex near São Paulo, Brazil (Domingos *et al.*, 1995). Similarly, Gordon *et al.* (1995) identified impacts of lead and zinc on a lichen species growing in the El Avila National Park, close to Caracas, Venezuela. In India, pollution is an increasing concern for important biological reserves, such as in the Dehra Dun–Rishikesh–Haridwar Valley of Uttar Pradesh (Gupta and Sharma, 1995).

It is encouraging to note that a number of initiatives have been taken to reduce transboundary air pollution problems in several of the regions covered by this book (Chapter 12). International cooperation has helped to solve or at least reduce transboundary air pollution problems in North America and Europe. The adoption of similar conventions could help to resolve many of the outstanding industrial pollution problems. The issue of smoke remains one that will be more difficult to solve. However, recognition of the extent of the problem (e.g. Haron *et al.*, 1997) is a first step in controlling the pollution.

References

Ademoroti, C.M.A. (1986) Levels of heavy metals on bark and fruit trees in Benin City, Nigeria. *Environmental Pollution*, Series B, 11, 241–253.

Alok, K. and Tewary, B.K. (1997) Moss (*Funaria hygrometrica*) as bioindicators for air pollution in Jharia Coalfield. *Journal of Ecobiology* 9, 145–156.

Andreae, M.O. (1991) Biomass burning: its history, use, and distribution and its impact on environmental quality and global climate. In: Levine, J.S. (ed.) *Global Biomass Burning. Atmospheric, Climatic and Biospheric Implications.* MIT Press, Cambridge, Massachusetts, pp. 3–21.

Banerjee, S.K., Nath, V., Bhowmik, A.K. and Manjhi, R.B. (1999) Soil characteristics and vegetation distribution in and around a lime kiln in Madhya Pradesh. *Annals of Forestry* 7, 77–86.

Beg, M.A.A. and Shams, Z.I. (1989) Impact of air pollution on vegetative growth (*Guaiacum officinale*) in Karachi. *Pakistan Journal of Scientific and Industrial Research* 32, 395–404.

Ben-Abdallah, F. and Boukhris, M. (1990) Action des pollutants atmosphériques sur la vegetation de la region de Sfax (Tunisie). *Pollution Atmosphérique* 127, 292–297.

Botha, A.T., Visser, J.H. and Moore, L.D. (1989) Evaluation of possible fluoride injury to vegetation in the vicinity of an industrial site near Cape Town. *South African Journal of Science* 85, 741–745.

Brighigna, L., Ravanelli, M., Minelli, A. and Ercoli, L. (1997) The use of an epiphyte (*Tillandsia caput-medusae* Morren) as bioindicator of air pollution in Costa Rica. *Science of the Total Environment* 198, 175–180.

Canas, M.S., Carreras, H.A., Orellana, L. and Pignata, M.L. (1997) Correlation between environmental conditions and foliar chemical parameters in *Ligustrum lucidum* Ait. exposed to urban air pollutants. *Journal of Environmental Management* 49, 167–181.

Carlson, C.A. (1997) Do concentrations of tropospheric ozone in eastern Mpumulanga pose a risk to commercial forests? *South African Journal of Science* 93, 7–10.

Carreras, H.A., Gudino, G.L. and Pignata, M.L. (1995) Comparative biomonitoring of atmospheric quality in five zones of Cordoba city (Argentina) employing the transplanted lichen *Usnea* sp. *Environmental Pollution* 103, 317–325.

Carreras, H.A., Canas, M.S. and Pignata, M.L. (1996) Differences in responses to urban air pollutants by *Ligustrum lucidum* Ait. and *Ligustrum lucidum* Ait. F. *tricolor* (Rehd.) Rehd. *Environmental Pollution* 93, 211–218.

Dhital, K. (1994) An overview of Nepal's fragile environment and development. *Economic Journal of Nepal* 17, 235–241.

Domingos, M., Poggiani, F., de Vuono, Y.S. and Silveira Lopes, M.I.M. (1995) Precipitação pluvial e fluxo de nutrients na floresta de Reservea Biologica de Paranapiacaba, sujeita aos poluentes atmosfericos de Cubatão, SP. *Revista Brasiliera de Botanica* 18, 119–131.

Douguedroit, A. and Bart, F. (1997) Winter dust haze in tropical southern Africa. *International Journal of Climatology* 17, 725–744.

El-Desouky, H.I., Moussa, K.F. and Hassona, H.H. (1998) Impact of automobile exhaust on roadside soils and plants in Sharkiya Governorate. *Egyptian Journal of Soil Science* 38, 137–151.

Ellegard, A. and Egneus, H. (1992) *Health Effects of Charcoal and Woodfuel Use in Low-income Households in Lusaka, Zambia.* Energy, Environment and Development Series, No. 14. Stockholm Environment Institute, Stockholm.

Ernst, W.H.O. (1985) Schwermetallimmissionen – ökophysiologische und populationsgenetische Aspekte. *Dusseldorfer Geobotanische Kolloquien* 2, 43–57.

Fang, M., Zheng, M., Wang, F., To, K.L., Jaafar, A.B. and Tong, S.L. (1999) The solvent-extractable organic compounds in the Indonesia biomass burning aerosols – characterization studies. *Atmospheric Environment* 33, 783–795.

Fatoki, O.S. (1987) Colorimetric determination of lead in tree leaves as indicators of atmospheric pollution. *Environment International* 13, 369–373.

Florig, H.K. (1997) China's air pollution risks. *Environmental Science and Technology* 31, 274A–279A.

Gonzalez, C.M. and Pignata, M.L. (1997) Chemical response of the lichen *Punctelia subrudecta* (NYL.) Krog transplanted close to a power station in an urban-industrial environment. *Environmental Pollution* 97, 195–203.

Gordon, C.A., Herrera, R. and Hutchinson, T.C. (1995) The use of a common epiphytic lichen as a bioindicator of atmospheric inputs to two Venezuelan cloud forests. *Journal of Tropical Ecology* 11, 1–26.

Granier, C., Hao, W.M., Brasseur, G. and Müller, J. F. (1997) Land-use practices and biomass burning: impact on the chemical composition of the atmosphere. In: Levine, J.S. (ed.) *Biomass Burning and Global Change.* MIT Press, Cambridge, Massachusetts, pp. 140–148.

Gupta, H.O. and Sharma, B.M. (1995) Impact of industrial acitivity on the ambient air quality in DehraDun–Rishikesh–Haridwar Valley (U.P.). *Indian Journal of Forestry* 18, 26–34.

Haron, A.H., Dahlan, T., Mohd. Puat, D. and Amran, M. (eds) (1997) *Transboundary Pollution and the Sustainability of Tropical Forests.* ASEAN Institute of Forest Management, Kuala Lumpur.

Helas, G., Bingemer, H. and Andreae, M.O. (1991) Organic acids over Equatorial Africa: results from DECAFE 88. *Journal of Geophysical Research* 97, 6187–6193.

Hyun, P. (1997) Evaluation of air pollution effects in Seoul City on forest soil at Mt Namsan by assay of denitrifying and sulfur-reducing bacteria. *Journal of the Korean Forestry Society* 86, 98–104.

Ichikawa, Y., Hatami, H. and Fujita, S.-I. (1997) Development of long-range transport models for East Asia. In: Kohno, Y. (ed.) *Proceedings of CRIEPI International Seminar on Transport and Effects of Acidic Substances.* Central Research Institute of the Electric Power Industry, Tokyo, pp. 67–76.

Innes, J.L., Beniston, M. and Verstraete, M.M. (eds) (1999) *Biomass Burning and its Inter-Relationships with the Climate System.* Kluwer Academic Publishers, Dordrecht.

Iqbal, M.Z. and Iqbal, H. (1995) Impact of dust from the factory at Karachi on the vegetation and soil characteristics. *Research Bulletin of the Punjab University of Science* 45, 73–78.

Iqbal, M.Z., Shafiq, M. and Rizvi, S.W.A. (1997) Effect of traffic exhaust on roadside trees during different seasons. *Polish Journal of Environmental Studies* 6(5), 55–59.

James, V. (1991) The Bhopal disaster: its impact on strategies for future transfers of hazardous technology. In: James, V. (ed.) *Urban and Rural Development in Third World Countries: Problems of Population in Developing Nations.* McFarland and Company Inc., Jefferson, North Carolina, pp. 113–133.

Kakulu, S.E. (1993) Biological monitoring of atmospheric trace metal deposition in North-Eastern Nigeria. *Environmental Monitoring and Assessment* 28, 137–143.

Khan, A.U., Siddique, S. and Naz, F. (1995) Effect of automobile exhaust on some tree species lining the Lahore Mall: a case study. *Pakistan Journal of Forestry* 45, 45–51.

Kharuk, V.I. (1999) Air pollution impacts on subarctic forests at Noril'sk, Siberia. In: Innes, J.L. and Oleksyn, J. (eds) *Forest Dynamics in Heavily Polluted Regions.* CAB International, Wallingford, pp. 77–86.

Kirchhoff, V.W.J.H. (1996) Increasing concentrations of CO and O_3: rising deforestation rates and increasing tropospheric carbon monoxide and ozone in Amazonia. *Environmental Science and Pollution Research* 3, 210–212.

Kirchhoff, V.W.J.H. and Pavão, H.G. (1997) The biomass burning sequence of the Brazilian cerrado and observations of atmospheric ozone. In: Levine, J.S. (ed.) *Biomass Burning and Global Change.* MIT Press, Cambridge, Massachusetts, pp. 599–605.

Klumpp, A., Domingos, M. and Klumpp, G. (1996a) Assessment of the vegetation risk by fluoride emissions from fertiliser industries at Cubatão, Brazil. *Science of the Total Environment* 192, 219–228.

Klumpp, A., Klumpp, G., Domingos, M. and da Silva, M.D. (1996b) Fluoride impact on native tree species in the Atlantic forest near Cubatão, Brazil. *Water, Air, and Soil Pollution* 87, 57–71.

Klumpp, A., Modesto, I.F., Domingos, M. and Klumpp, G. (1997) Susceptibility of various *Gladiolus* cultivars to fluoride pollution and their susceptibility for bioindication. *Pesquisa Agropecuaria Brasileira* 32, 239–247.

Klumpp, A., Domingos, M., de Moraes, R.M. and Klumpp, G. (1998) Effects of complex air pollution on tree species of the Atlantic rain forest near Cubatão, Brazil. *Chemosphere* 36, 989–994.

Lacaux, J.-P., Delams, R.A., Cros, B., Lefeivre, B. and Andreae, M.O. (1991) Influence of biomass burning emissions on precipitation chemistry in the Equatorial forests of Africa. In: Levine, J.S. (ed.) *Global Biomass Burning. Atmospheric, Climatic and Biospheric Implications.* MIT Press, Cambridge, Massachusetts, pp. 167–173.

Lee, C.K., Kim, J.K. and Cho, K.S. (1998) Studies on the relation between tree injury and acid precipitation. *Korean Journal of Environment and Ecology* 12, 131–137.

Lee, K.J., Cho, W. and Han, B.H. (1995) The effect on the forest plant community by the air pollution around Daesan and Kunsan industrial complex areas. *Journal of Korean Applied Ecology* 9, 1–14.

Lee, S.W., Woo, S.Y., Koo, Y.B. and Lee, S.K. (1998) Genetic differences between the tolerant and sensitive trees in an air polluted *Prunus sargentii* stand. *Journal of the Korean Forestry Society* 87, 74–81.

Levine, J.S. (ed.) (1996) *Biomass Burning and Global Change.* MIT Press, Cambridge, Massachusetts.

Liang, Y., Izuta, T., Totsuka, T., Yokoyama, M. and Ma, L.Q. (1994) Effects of sulfur dioxide on growth, gas exchange rate and leaf sulfur content of Masson pine seedlings. *Journal of Environmental Sciences* 6, 268–276.

Liu, H.T., Zhang, W.P., Shen, Y.W., Du, X.M., Zhou, X.Y., Liu, R.A., Ma, L.Q., Du, S.C. and Tang, Q. (1988a) Relationship between acid rain and the decline of masson pine forest in Nanshan, Chongqing. *Acta Scientiae Circumstantiae* 8, 331–339.

Liu, H.T., Zhang, W.P., Shen, Y.W., Du, X.M., Zhou, X.Y., Liu, R.A., Ma, L.Q., Du, S.C., Lan, K.H., Lou, R., Lou, J.J. and Tang, Q. (1988b) A study on some symptoms of the declining and dying *Pinus massoniana* Lamb. Forest on Nanshan Mountain in Chongqing. *Acta Botanica Sinica* 30, 318–324.

Liu, H.Y., Chang, W.L., Oltmans, S.J., Chan, L.Y. and Harris, J.M. (1999) On springtime high ozone events in the lower troposphere from Southeast Asian biomass burning. *Atmospheric Environment* 33, 2403–2410.

Maggs, R., Wahid, A., Shamsi, S.R.A. and Ashmore, M.R. (1995) Effects of ambient air pollution on wheat and rice yield in Pakistan. *Water, Air, and Soil Pollution* 85, 1311–1316.

Morales, J.A., Bifano, C. and Escalona, A. (1995) Rainwater chemistry at the western Savannah region of the Lake Maracaibo Basin, Venezuela. *Water, Air, and Soil Pollution* 85, 2325–2330.

Mou, S.S. and Qing, C.L. (1995) Mercury pollution of soil-crop system in acid precipitation area. *Pedosphere* 5, 283–288.

Muaka, M., Komanda, A. and Mutomb, M. (1989) Quelques considerations sur la pollution de l'air à Lubumbashi (Shaba, Zaire). *Geo-Eco-Trop* 13, 113–125.

Nagaya, Y., Taniyama, T. and Yoshizaki, S. (1998) Relationship between distribution of air pollutants and vital degree (vitality) of trees around Kwangyang Bay in Korea. *Bulletin of the Faculty of Bioresources, Mie University*, 1998(20), 1–9.

Nyangababo, J.T. and Salmeen, A. (1987) Deposition of airborne elements from Sembule steel mill surveyed by moss analysis. *International Journal of Environmental Studies* 30, 177–183.

Okazaki, M., Xu, G.S. and Sun, Z.P. (1998) Effects of acidic deposition on soil ecosystems under forest in the Chongqing region of China. *Soil Science and Plant Nutrition* 44, 187–196.

Panes, V.A. and Zamora, P.M. (1991) Leaf epidermal features of four Philippine plants as indicators of cement dust pollution. *Philippine Journal of Science* 120, 249–267.

Park, J.W. and Cho, S.Y. (1998) A long range transport of SO_2 and sulfate between Korea and East China. *Atmospheric Environment* 32, 2745–2756.

Quintero, S. and Ramirez, J. (1990) El ozono, agente etiologico de la necrosis ambiental del tabaco. *Agrotécnica de Cuba* 22, 41–49.

Radojevic, M. and Hassan, H. (1999) Air quality in Brunei Darussalam during the 1998 haze episode. *Atmospheric Environment* 33, 3651–3658.

Rajput, M.S., Kanwar, J.S. and Bajwa, M.S. (1971) Control black tip of mango with caustic soda. *The Punjab Horticultural Journal* 11, 49–51.

Ram, S. (1989) Factors associated with black tip and internal necrosis in mango and their control. *Acta Horticulturae* 231, 797–804.

Rutllant, J. and Garreaud, R. (1995) Meteorological air pollution potential for Santiago, Chile: towards an objective episode forecasting. *Environmental Modelling and Assessment* 34, 223–244.

Salgare, S.A. and Trisa, P. (1997) Effect of ambient air quality at Mumbai on pollen production of some angiospermic plants. *Journal of Ecotoxicology and Environmental Monitoring* 7(2), 155–156.

Schmitt, S.F. and Whittaker, R.J. (1998) Disturbance and succession on the Krakatau Islands, Indonesia. In: Newbery, D.M., Prins, H.H.T. and Brown, N.D. (eds) *Dynamics of Tropical Communities*. Blackwell Science, Oxford, pp. 515–548.

Sharma, C.K. (1996) Overview of Nepal's energy sources and environment. *Atmospheric Environment* 30, 2717–2720.

Shen, J., Zhao, Q., Tang, H., Zhang, F., Feng, Z., Okita, T., Ogura, N. and Totsuka, T. (1995) Concentrations and depositions of SO_2, SO_4^{2-} etc. in a Chongqing suburban forested area. *Water, Air, and Soil Pollution* 85, 1299–1304.

Song, K.J., Choi, Y.C. and Lee, K.J. (1996) Acid rain and airborne pollutants effects on the needles of some conifer species – a case study of injury index and contact angles. *Korean Journal of Environment and Ecology* 10, 76–86.

Subido, P.S., Santamaria, P.A., Alqueza, E.N. and Pizarro, A.C. (1969) Plants injured by air pollutants. *Philippine Journal of Science* 98, 323–328.

Suckcharoen, S. (1980) Mercury contamination of terrestrial vegetation near a caustic soda factory in Thailand. *Bulletin of Environmental Contamination and Toxicology* 24, 463–466.

Verma, N.K., Pandey, D.D. and Upadhyay, S.K. (1992) Effect of coal dust pollution on biomass, primary productivity and nutritive value of *Aristida cyanatha* dominated grasslands of Bihar. *Journal of Applied Biology* 2, 87–95.

Vijayan, R. and Bedi, S.J. (1989) Effect of chlorine pollution on three fruit tree species at Ranoli near Baroda, India. *Environmental Pollution* 57, 97–102.

Wakamatsu, S., Utsunomiya, A., Han, J.S., Mori, A., Uno, I. and Uehara, K. (1996) Seasonal variation in atmospheric aerosols concentration covering northern Kyushu, Japan and Seoul, Korea. *Atmospheric Environment* 30, 2343–2354.

Wu, G. (1998) Effect of acidic deposition on the productivity of the forest ecosystem and estimation of its economic losses in the southern suburbs of Chongqing, China. *Journal of Environmental Sciences* 10, 210–215.

Yu, S.W., Yu, Z.W., Ma, G.J., Zhu, C.L., Bian, Y.M., Chen, S.Y. and Zhu, X.Y. (1988) Preliminary investigation on the causes of masson pine forest decline in Nanshan region of Chongqing, Sichuan Province. *Chinese Journal of Environmental Science* 9, 77–81.

Yu, S.W., Bian, Y.M., Ma, G.J. and Luo, J.J. (1990a) Studies on the causes of forest decline in Nanshan, Chongqing. *Environmental Monitoring and Assessment* 14, 239–246.

Yu, S.W., Bian, Y.M., Ma, G.J. and Luo, J.J. (1990b) Decline of Masson pine forest in Nanshan, Chongqing and air pollution. *Acta Scientiae Circumstantiae* 10, 378–383.

Zhang, F.Z., Zhang, J.Y., Zhang, H.R., Ogura, N. and Ushikubo, A. (1996) Chemical composition of precipitation in a forest area of Chongqing, southwest China. *Water, Air, and Soil Pollution* 90, 407–415.

Zhao, D.W. and Seip, H.M. (1991) Assessing effects of acid deposition in southwestern China using the MAGIC model. *Water, Air, and Soil Pollution* 60, 83–97.

Zheng, Y., Stevenson, K.J., Barrowcliffe, R., Chen, S., Wang, H. and Barnes, J.D. (1998) Ozone levels in Chongqing: a potential threat to crop plants commonly grown in the region. *Environmental Pollution* 99, 299–308.

General Problems Associated with Air Quality in Developing Countries

2

N. Abdul Rahim

Forest Research Institute Malaysia, Kuala Lumpur, Malaysia

Many developing countries have experienced a progressive degradation in air quality as a result of their rapid pace of development over the last two decades. The main factors causing the deterioration include rapid urbanization, an increase in the number of motor vehicles and rising consumption of energy. Other important causes include agricultural intensification, refuse disposal and household heating. These factors have not only influenced environmental quality, but also affect human health. The air pollution problems in South-east Asia are especially highlighted in view of the impressive economic growth in the region during the 1980s and 1990s and the rapid increase in the population of the area. The effects of the haze episodes on air quality and health that enveloped the region in 1997 are analysed, including the economic cost of the catastrophe.

1 Introduction

Rapid development over the last two decades has seen a progressive deterioration in air quality in many developing countries. This has occurred because, during that period, many developing and newly industrialized countries underwent or experienced unparalleled economic growth, swelling urban populations and generated excessive emissions from automobiles, factories and refuse burning (World Bank, 1996). While economic growth raises living standards and provides people with the means to enjoy their environment, development can quite often lead to severe pressure on natural resources and the environment. This question has attracted lengthy discussion at many international fora, including the United Nations Conference on Environment and Development in 1992, and has been addressed by international conventions such as the United Nations Framework Convention on Climate Change.

Although the available evidence clearly shows that air quality problems are primarily confined to urban areas and cities, it is likely that poor air quality will become increasingly important in rural areas as countries forge ahead with rapid development. This is illustrated by the haze episodes that enveloped large areas of South-east Asia at the end of 1997 (Glover and Jessup, 1999). In many towns and cities of the world, exposure to air pollution is the main environmental threat to human health. The three most important pollutants affecting human health in developing countries are lead, dust and soot (World Resources Institute, 1998). In developing countries, more than half of the gross domestic product comes from cities. As the process of urbanization has accelerated, cities in Africa, Asia and Latin America have collided with the challenges of congestion and pollution (UNEP, 1997).

This chapter reviews and highlights trends in air quality in developing countries, with emphasis on South-east Asian countries. The occurrence of haze episodes in this region is also discussed. Further details of the haze episodes can be found in Chapters 8 and 11.

2 Trends in urbanization

Much of the 20th century witnessed an increasing trend in urbanization in developing countries. While urbanization can be a stimulus of development, in the process many cities in Africa, Asia, the Near East and Latin America are facing the twin challenges of pollution and congestion (Table 2.1). In addition, growing urban populations in low- and middle-income countries have become an inescapable phenomenon. While the rate of total population growth and urbanization in developing countries is increasing, rates in developed countries are slowing down. In 1960, less than 22% of the developing world's population was urban and the proportion increased to 34% by 1990. The World Bank has estimated that 50% of the global population will be urban by 2020. As urban populations rise toward extreme levels, the provision of basic services deteriorates and air pollution problems develop. This trend is clearly evident in Mexico City (see Chapter 3). In Asia, the urban population is expected to reach 2.5 billion by 2025, three times what it was in 1990 (Brandon, 1998). At present, 87 Asian cities have more than 1 million inhabitants.

Urbanization has significantly contributed to the economic growth and better standard of living in many developing countries. However, a major concern is the sustainability of the urban systems, not only as a suitable human habitat but also in relation to ecological and environmental support systems (Sham Sani, 1994). Evidence thus far indicates that the general state of the

Table 2.1. Air pollution in selected developing, rapidly industrializing and transition countries. (Source: World Resources Institute, 1994.)

Country	City	City population (× 1000)	Total suspended particulates (μg m^{-3})	Sulphur dioxide (μg m^{-3})	Nitrogen dioxide (μg m^{-3})
Brazil	São Paulo	16,533	86	43	83
Chile	Santiago	4,891	–	29	81
China	Shanghai	13,584	246	53	73
Colombia	Bogota	6,079	120	–	–
Croatia	Zagreb	981	71	31	–
Ecuador	Guayaquil	1,831	127	15	–
	Quito	1,298	175	31	–
Egypt	Cairo	9,690	–	69	–
Ghana	Accra	1,673	137	–	–
India	Mumbai	15, 138	240	33	39
Indonesia	Jakarta	8,621	271	–	–
Iran	Tehran	6,836	248	209	–
Kenya	Nairobi	1,810	69	–	–
Korea, South	Seoul	11,609	84	44	60
Malaysia	Kuala Lumpur	1,238	85	24	–
Mexico	Mexico City	16,562	279	74	130
Philippines	Manila	9,286	200	33	–
Poland	Katowice	3,552	–	83	79
Singapore	Singapore	2,848	–	20	30
South Africa	Cape Town	2,671	–	21	72
	Johannesburg	1,849	–	19	31
	Durban	1,149	–	31	–
Thailand	Bangkok	6,547	223	11	23
Turkey	Istanbul	7,911	–	120	–
Ukraine	Kiev	2,809	100	14	51
Venezuela	Caracas	3,007	53	33	57

environment, including air quality, is deteriorating in many cities. World Bank studies in selected cities of developing countries have demonstrated that swelling urban populations and the growth of industrial activities and automotive traffic in Asia have caused severe air pollution (Shah and Nagpal, 1997).

According to a World Resources Institute (WRI) assessment made in 1998, more than 600 million people live in urban areas where sulphur dioxide levels exceed the guidelines of the World Health Organization (WHO) and over 1.25 billion live in cities with unacceptable levels of suspended particulate matter (World Resources Institute, 1998). Another study, conducted jointly by the United Nations Environment Programme (UNEP) and WHO, and covering 20 of the world's megacities, concluded that the situation in these cities was alarming (UNEP/WHO, 1992). The study noted that every city had at least one major pollutant that exceeded WHO health guidelines, 14 cities had at least two and seven cities had at least three. The World Development Indicators 1998 also highlighted the severe air pollution problems present in many developing countries (World Bank, 1998).

3 Trends in air pollution

Air pollution is normally defined as air that contains one or more chemicals or possesses a physical condition that may harm humans, animals or plant species (Miller, 1985). Air pollution resulting from human activities is not new. In England, the first known air quality laws were declared in 1273 by King Edward I, aimed at controlling sulphur. In the USA, the industrial revolution brought severe air pollution to many areas, mainly from coal-mining industries. However, as developing countries make economic progress, they risk emulating the mistakes of the past by placing too much emphasis on growth while neglecting the environment. Very often, economic growth in developing countries is accompanied by urbanization, more motor vehicles and increased energy consumption. Such development places heavy pressure on a country's environmental resources.

Air pollution does not respect boundaries. It affects landscapes and ecosystems far from its sources, as experienced by those lakes in Scandinavia that were adversely affected by transboundary pollution. As a result, in 1979 the United Nations Economic Commission for Europe (UN-ECE) organized the Convention on Long-Range Transboundary Air Pollution as a means to address the problem. The approach adopted by the signatories to this convention is described by Bull and Fenech (1999).

With rapid industrialization and increased energy use, air quality deterioration, which at one time only affected industrialized countries, is now apparent in many developing countries in the Asia-Pacific region and in Latin America. The UNEP study *Global Environment Outlook* noted the significant increase in air pollution in developing countries, particularly Asia-Pacific (UNEP, 1997). Recognizing the growing seriousness of air pollution in some cities, the World Bank started the Urban Air Quality Management Strategy (URBAIR) in 1992, covering large cities such as Mumbai (India), Jakarta (Indonesia), Kathmandu (Nepal), Metro Manila (Philippines) and Colombo (Sri Lanka) (World Bank, 1997a,b).

The major pollutants produced by human activities include sulphur oxides, carbon monoxide, nitrogen oxides and particulate matter. These pollutants account for 90% of the air pollution problems in the USA. The situation is rather different in developing countries, where the primary pollutants are suspended particulate matter, sulphur dioxide, nitrogen oxides and carbon monoxide (World Resources Institute, 1994) (Table 2.1). Of late, haze episodes have occurred in some countries in Asia and the Pacific. However, air pollution monitoring in developing countries is still inadequate or at best rudimentary, making it difficult to describe an overall picture. Worldwide, an estimated 1.1 billion urban residents are exposed to suspended particulate matter or sulphur dioxide levels in excess of the guidelines set by WHO (UNEP, 1997).

Understandably, major differences exist between regions in the approach to environmental issues, reflecting differences in priorities and geographical factors. Acid rain and transboundary air pollution are now becoming apparent in parts of Asia and the Pacific region as well as in parts of Latin America and Africa. Regions with emerging economies, such as Eastern Europe, South Asia and parts of Latin America, face problems associated with rapid industrialization. Rising levels of air pollution have caused a deterioration in urban air quality and are creating acidification problems that ultimately pose a health risk. The *Global Environment Outlook* published by UNEP in 1997 noted that air pollution is considered to be critically important in Asia-Pacific and Latin America as well as in North America and Europe. However, whereas air quality is deteriorating in Asia-Pacific and Latin America, both Europe and North America are showing signs of improved air quality, at least for some pollutants. Within the Asian and Pacific regions, industrial pollution is a high priority in only Pakistan, Indonesia, Malaysia and the Philippines.

Thus, air quality and the acidification of the environment have begun to emerge as significant issues in parts of the Asia and Pacific regions. It has been estimated that 31 million tonnes (Mt) of SO_2 were emitted in the 22 countries of Asia in 1990, almost 50% more than in North America (Shrestha *et al.*, 1996). About 78% of these emissions originated from outside East Asia. Another study of acid rain and emissions of acidifying pollutants in Asia has shown that there are areas in southern China, south-east Thailand, Cambodia and southern Vietnam where critical loads are being exceeded, with up to 320 mg acidity m^{-2} $year^{-1}$ being recorded (Hettelingh *et al.*, 1995).

4 Common sources of air pollution in developing countries

The causes of air pollution problems vary considerably between countries. While much of the increasing severity of the problem is caused by the demographic situation in individual countries, other aspects of human conditions – such as changing patterns of consumption – have played a significant role. In this respect there are three primary sources of air pollution in the developing world: transportation, industrialization and energy generation (World Resources Institute, 1996). The importance of each differs between countries. For example, in the Philippines, the main contributor to air pollution is transport, followed by power generation, industrial units and the burning of refuse. Other important sources include agricultural intensification, refuse disposal, and household heating and cooking. As a result of agricultural and forest conversion activities, forest fires have increasingly become a dominant and damaging cause of air quality problems in countries such as Brazil, Indonesia, Brunei Darussalam and Malaysia (Glover and Jessup, 1999). The fires are of sufficient importance to have been dealt with separately (Chapter 11).

4.1 Transportation and motor vehicles

In many countries, by far the largest amounts of pollutants are emitted by motor vehicles. This is partly because of the relatively high densities of road networks that have been built in the cities of developing countries over the last two decades and also because of the increase in the number of motor vehicles (associated with both the increasing populations of the cities and an increase in the *per capita* numbers of vehicles). These developments have resulted in rapidly deteriorating air quality in many cities. One of the best documented cases is Mexico City (Chapter 3). Motor vehicles account for the major share of pollutant emissions in Delhi (57%), Beijing (75%), Manila (70%) and Kuala Lumpur (86%) (World Resources Institute, 1994). Similarly, Latin America is not spared from increasing problems associated with air quality, as 78% of the region's population lives in cities (United Nations, 1995). The two most important factors causing an increase in urban air pollution are the increasing numbers of motor vehicles in use and the expansion of industrial activities. Mexico City alone has more than 4.2 million motor vehicles (Cevalos, 1996), while in Santiago de Chile the number has tripled in recent years. The use of motor vehicles is responsible for generating 80–90% of atmospheric lead concentrations, with the combustion of leaded gasoline being the source. Mexico City regularly experiences the worst air quality in the world, a situation that arises from the combination of the large numbers of vehicles and the topographic and climatic situation (Chapter 3). Concentrations of sulphur dioxide, suspended particulate matter and carbon monoxide in the cities of developing countries exceeded WHO guidelines during the 1970s and 1980s (UNEP/WHO, 1992), mainly as a result of emissions from vehicles. However, in some cities, there has been a marked improvement over the past 5 years. Motor vehicles, primarily passenger cars, are also the largest source of hydrocarbons and nitrogen oxides in city areas. In the urban area of Kuala Lumpur and Petaling Jaya, Malaysia, motor vehicles account for some 92% of total pollutant emissions, with 97% of these emissions comprising carbon monoxide, hydrocarbons and nitrogen oxides (Sham Sani, 1987, 1994).

4.2 Industries

Industrial air pollution has been regarded as a problem restricted to the western world but this is unfortunately no longer true. Following the rapid growth of industries and intensified economic development, industrial air pollution is now a feature of a number of developing countries, as shown by the contributions to this book. Although the sizes of industrial plants in these countries are comparatively small by western standards, it is important that planners be aware of the cumulative effect of these many small industrial sources of pollution. Indeed such individual sources may contribute more to area problems than single large sources such as power plants or refineries.

In many of the rapidly developing countries of the ASEAN (Association of South-East Asian Nations) region, industrialization occurred at triple the pace of the industrial revolution of the west (ASEAN, 1997). Consequently, many parts of the region are experiencing industrial pollution on a scale not seen in the developed world for the last four decades. Air pollution problems are particularly acute in cities such as Rio de Janeiro (Brazil), Jakarta (Indonesia) and Bangkok (Thailand).

A problem that needs to be addressed is the displacement of polluting industries to developing countries. The improvement of air quality in Europe and North America has occurred through increasingly strict regulation of emissions. However, the costs of such controls are increasing on a per unit emission basis. As a result, there is a risk that some industrial plants will be relocated to areas where less emphasis is placed on air quality. Other priorities in such regions, such as regional development, may result in pollution being seen as a trade-off for the economic benefits that the industry will bring. Such a trend is contrary to the concept of sustainable development, and both developed and developing countries need to devise mechanisms to prevent such a situation arising.

4.3 Power plants

As a result of economic growth in the Asian and Pacific countries, energy demand has increased remarkably. In 1992, the region, excluding Japan, Australia and New Zealand, accounted for 21% of the world's primary commercial energy demand, compared with 51% from OECD countries and 28% from the rest of the world (UNEP, 1997). The growth in energy demand for the whole region was 3.6% $year^{-1}$ between 1990 and 1992 compared with an average of 0.1% growth for the whole world (ADB, 1994). At the same time, the region accounted for about 41% of world coal consumption in 1993 (EIA, 1995).

In many developing countries, large power plants are using fossil fuels. As sulphur is one of the major ingredients of these fuels, power plants are the greatest contributors of sulphur dioxide pollution. Fuel oil has an emission rate of 21.1 kg of carbon per gigajoule (GJ) compared with natural gas, which has an emission rate of 15.3 kg carbon GJ^{-1}. Because of this, a number of countries, particularly in the South-east Asian region, are now seriously considering the use of low-sulphur coal as an alternative source of energy for power stations. The environmental impact of such a switch is not expected to be immediate for the whole region or country, as power companies in respective countries may need some time to adjust or accommodate their plants' physical structures or requirements.

The over-reliance on coal translates into a significant increase in air pollution. Fly ash produced from coal burning is also a significant problem in the region, especially in India, where 35.4 Mt of fly ash are generated by power plants every year and only 2–3% is being reused (Government of India, 1993). The coal burning has resulted in acidic deposition in surrounding areas.

4.4 Forest conversion and agricultural activities

During the last two decades, a tremendous amount of land conversion has occurred in many developing countries, especially in Asia and Latin America. Huge tracts of tropical forest have been converted to other land use practices such as agriculture, grazing, urban development and industrial expansion, which normally bring much higher returns. The FAO (1997) reported that between 1980 and 1995, the developing world lost nearly 200 Mha of natural forests, mostly through clearing for agriculture, which includes shifting cultivation, cash crops, ranching and other subsistence agriculture. The greatest losses in the tropical zone of developing countries were in tropical Asia–Oceania (0.98% $year^{-1}$). Unfortunately, the losses have only been partially compensated for by new forest plantations.

Widespread use of fire to clear previously logged-over forests in order to establish grazing areas or large plantations of oil palms (*Elaeis guineensis*), rubber trees (*Hevea brasiliensis*) or pulpwood has reduced air quality. In some instances, land is also cleared by fire for slash-and-burn practices in shifting cultivation. As a result of such activities, reduced air quality is not only a characteristic of cities, but also a feature of some rural areas. Improper management of agricultural land, especially in the semiarid regions of the world, results in loss of vegetative cover and subsequent soil erosion (Goldammer, 1997). The burning of fields creates smoke and suspended particulates. Occasionally, as a result of drought conditions and improper management, and sometimes because of deliberate acts, the fires spread to secondary and primary forest areas, grassland and peat bogs. The smoke from such fires can extend across large areas and, when combined with urban air pollution, may cause severe health problems and extensive social and economic damage.

4.5 Refuse disposal

Population growth has had a dramatic impact both on the amount of solid waste produced in urban areas and on the demand for municipal collection and disposal systems. The amount of refuse produced in large cities of the world is expected to increase, not only as the population rises, but also in terms of kilograms of refuse generated per person per day.

Open burning of dumps and refuse piles is a very inefficient form of combustion and releases a great deal of smoke and particulate matter into the atmosphere. However, being the easiest and most convenient method of disposal, open burning is practised almost everywhere, especially in developing countries. Individually, open-burning sources may not appear to be significant, but because of their numbers, this method of waste disposal contributes substantially to reduced air quality.

Tipping (landfill) is another commonly used method of refuse disposal. Unlike open burning, disposal by tipping, if well planned and controlled, presents not only an economic answer to the disposal problem, but in certain cases it can also fulfil a social need for the reclamation of certain areas to provide playing fields, parkland and, in some cases, sites for industrial developments and housing areas. However, where landfill operations are not carefully controlled, odours from partly decomposed refuse can become very offensive in the neighbourhood of the tip and inadequate compaction and covering of refuse can lead to infestations by flies and vermin. In addition, the controlled burning of methane generated during the decomposition of the refuse is important, and methane may even be harnessed as a source of energy.

4.6 Household heating and cooking

Smoke from household heating and cooking is another important source of pollution in developing countries (see Chapter 10). Household fuel choice in developing countries can be seen as an energy ladder, with dirty fuels such as crop residues and firewood at the bottom. These are followed by charcoal, kerosene and liquid gas. Charcoal and coal are fairly common in urban areas while wood and crop wastes are commonly used in rural and poor or middle-class households. Emissions from coal generally depend on the type of coal, but they can be relatively high in both particulate matter and carbon monoxide.

5 Forest fires and haze episodes in South-east Asia

Since the early 1980s, haze episodes have been observed in some parts of Malaysia and Indonesia (Sham Sani *et al.*, 1991). These episodes originate from the widespread use of fire to clear previously logged forest and other degraded land in preparation for oil palm, rubber, or pulpwood plantations and, sometimes, rice in peat areas. To a lesser extent, land is also cleared by fire for slash-and-burn agriculture. In the process, a series of major forest fires have been initiated, causing significant environmental degradation in the region.

Fires were particularly apparent in 1990, 1991, 1994 and 1997, with the most severe occurring in October 1997 (Leong and Lim, 1997). The impacts of such forest fires have caused haze phenomena that have enveloped many areas of Indonesia, Malaysia and Singapore (Ayers *et al.*, 1997). Similar disasters have been experienced in other regions, including Florida, Mexico, Brazil and Australia.

The haze events have prompted the National Haze Committee in Malaysia to coin a clear definition for haze, namely that it is a form of air pollution caused by the presence of large number of minute particles suspended in the atmosphere. These particles absorb and scatter light; in doing so, they reduce visibility and give air an opalescent or hazy appearance. Haze occurs whenever fine particulates from emissions are trapped in a stable atmosphere. In Malaysia, such conditions usually occur between July and September, during the south-west monsoon season, when there is a reduction in the conditional instability of the atmospheric circulation that would otherwise form a deep convection which would disperse and dilute the concentration of the pollutants in the atmosphere efficiently through convective mixing, causing widespread heavy rain to wash out the trapped pollutants (Leong and Lim, 1997). Generally, urban industrial areas contribute a fairly consistent pollutant load to the atmosphere that is moderated or enhanced by daily transportation peaks, the diurnal cycle and local weather conditions.

In the past two decades, Malaysia has experienced no fewer than five episodes of intense haze. These occurred during April 1983 (Chow and Lim, 1983), August 1990 (Sham *et al.*, 1991), June 1991 and October 1991 (Cheang *et al.*, 1991) and again in August–October 1994 (Malaysian Meteorological Service, 1995). In contrast with the shallow, localized, transient haze conditions occurring in urbanized areas, these dense haze episodes have been aggravated by the injection of suspended particulates from volcanic eruptions (Mount Pinatubo, June 1991) or large-scale forest and plantation fires. Forest fires normally inject a range of biomass combustion products into the atmosphere, including abundant aerosols, trace gases such as nitrous oxide (NO), carbon dioxide (CO_2), nitrogen oxides (NO_x), nitric oxide (N_2O), ammonia (NH_3), sulphur oxides (SO_x), methane (CH_4) and other non-methane hydrocarbons. These are picked up by the monsoonal winds, transported, and deposited along the path of the wind at considerable distances away from the source. The haze condition is also aggravated by smoke from localized open burning.

The worst haze event experienced in Malaysia, which affected a significant part of South-east Asia and received wide media coverage, was the 1997 haze 'episode'. During the September–November 1997 haze, Indonesia, Singapore, Brunei Darussalam and Malaysia were severely affected. While internal sources such as open burning and pollutant discharges from motor vehicles also contributed to the problem, the main cause of the haze was the forest and peat fires in Kalimantan and Sumatera, involving some 300,000 ha of land, and the meteorological conditions prevailing at the time. In the latter case, the south-west monsoon winds blew across Kalimantan and Sumatera transporting the pollutants from the fires to Singapore, Brunei Darussalam and almost all parts of Malaysia. The episode was further exacerbated by the simultaneous El Niño phenomenon that affected Indonesia and most of South-east Asia, prolonging the dry season and intensifying forest fire risks.

The air pollution index (API) during the haze period in September and October of 1997 for urban centres in the Klang Valley in Malaysia revealed that all centres experienced 'unhealthy' conditions for more than 50% of the time; Kuala Lumpur for 69.4% of the time, Petaling Jaya for 57.2%, Klang for 63.8% and Kajang for 54.0%. The proportion of 'very unhealthy' days was highest in Kuala Lumpur (20.4%), and lower at the other sites: 8.2% of days at Petaling Jaya, 4.2% at Klang and 4.0% at Kajang. Kuala Lumpur also recorded 2% of days with a 'hazardous' API reading. The El Niño event, which continued until mid-1998, prolonged the dry season in many parts of Malaysia, Indonesia and South-east Asia. In Malaysia, the prolonged dry season severely affected water supply in the Kuala Lumpur and Klang Valley areas. The situation became so critical that water rationing was imposed at the beginning of April 1998.

Following the 1997 haze episode, several initiatives were introduced in order to contain and prevent similar episodes at the national and regional levels. The international community and friendly countries around the world also contributed to the initiatives both during the height of the disaster and later in formulating preventive measures. Indonesia, Singapore and Malaysia agreed in December 1997 to take collective measures to prevent future haze due to forest and peat fires. Three objectives were identified: to prevent land and forest fires through better management policies and enforcement; to establish operational mechanisms to monitor land and forest fires; and to strengthen regional land and forest-fire fighting capability and other mitigating measures.

6 Air pollution problems in South-east Asian countries

The seven countries in South-east Asia that form the regional cooperative ASEAN, are Brunei Darussalam, Indonesia, the Philippines, Malaysia, Singapore, Thailand and Vietnam. The total land area of ASEAN member countries is 338.7 Mha, with a population of 406.6 million. Economic strategies adopted by ASEAN have influenced demographic dynamics and land use patterns. On the whole, member countries have presented a remarkable economic achievement during the last two decades. Most countries have attained an average annual growth rate of 10% or higher. At the same time, the population has been expanding very rapidly, almost doubling every 30 years. The urban population is projected to grow by 3.5% annually, outpacing the average growth of the population, which is estimated at 1.6% each year (ASEAN, 1997). Industrialization has severely affected the resource base and has resulted in

Table 2.2. Emissions of SO_2 (kg year^{-1}) from anthropogenic sources in the ASEAN region. (Source: ASEAN, 1997.)

Date	Total	Per capita
1975	1550	5.6
1980	2218	7.1
1985	1909	5.5
1990	1925	5.0
Change (1975–1990)	24%	

significant increases in energy use. As a result, air pollution can on occasion reach critical levels in highly urbanized or industrialized areas.

In 1990, per capita emissions of sulphur dioxide from man-made sources in the ASEAN region were only 5 kg year^{-1}, compared with a range of 9 kg to > 150 kg year^{-1} in developed countries (Table 2.2; ASEAN, 1997; Vancura *et al.*, 1999). Total emissions in ASEAN still show an increasing trend, in contrast to developed countries (such as USA and Japan) that have been able to reduce their total emissions. Per capita emissions in ASEAN, as in other countries, are decreasing.

To reduce sulphur dioxide emissions, ASEAN countries are shifting to the use of low-sulphur fuel (preferably less than 1%), particularly for industries near urban centres. This practice is being encouraged by stricter national emission standards.

6.1 Nitrogen oxides

Anthropogenic emissions of nitrogen oxides are mostly derived from power stations and vehicular emissions. Combustion produces several oxides of nitrogen. Those known to have adverse environmental or biological effects are NO and NO_2. Total nitrogen oxide emissions in ASEAN increased during the period 1980–1990, in contrast to decreasing levels in developed countries such as the USA and Japan (Table 2.3).

6.2 Factors affecting urban air quality in South-east Asia

The deterioration of air quality in ASEAN urban areas is mainly the result of industrial expansion and rapid population growth, and the increase in the number of vehicles. The transport sector, in particular, has been responsible for emissions of substantial quantities of lead, particulates, carbon monoxide, hydrocarbons and nitrogen dioxide. Today, three of the ASEAN capital

Table 2.3. Emissions of NO_x in the ASEAN region from anthropogenic sources (1000 t year^{-1}). (Source: ASEAN, 1997.)

Year	Total	Per capita
1975	940	3.4
1980	1189	3.8
1985	1412	4.0
1990	1582	4.1
Change (1975–1990)	33%	

cities rank among the world's 30 largest urban agglomerations in terms of population size. In the year 2000 the populations of Bangkok (Thailand), Jakarta (Indonesia) and Manila (Philippines) will each reach at least 10 million, qualifying them to be among the 20 megacities in the world.

Rapid population growth has increased the demand for transportation. The number of passenger vehicles on the road has grown over the years, and so has the demand for public transportation. However, the inability of some countries to cope with the transport requirements of rapid urbanization has exacerbated the problem of urban congestion. Inadequate public transport systems have made people want to own their own vehicles, thus aggravating traffic and adversely affecting the quality of roads. The resulting traffic congestion increases fuel consumption and the levels of pollutants emitted into the air.

Traffic congestion is forcing people to spend more time on the road and to be exposed for longer to higher levels of pollution. Emissions can adversely affect health, especially of the very young and very old members of the population. School children aged 14 years and below (about 30% of the population) are the most affected. This group of children are the most susceptible to deteriorating air quality because they spend more time on the road travelling to and from school and playing outside.

As ASEAN becomes more affluent, the number of motor vehicles will grow more rapidly. At present, the number of vehicles is increasing at an average annual rate of almost 8%. In 1990, more than 8% of the total number of vehicles in the ESCAP (Economic and Social Commission for Asia and the Pacific) region could be found in ASEAN (ESCAP, 1990, 1994). This number excludes the several million motorcycles and other vehicles with fewer than four wheels. Thailand, Indonesia and Malaysia account for more than 80% of the motor vehicles in the region.

The concentration of motor vehicles in cities makes their impact on the urban environment more pronounced. The ratio of cars per thousand population in ASEAN capitals ranges from 75 to 555 (Table 2.4). It is highest in Bandar Seri Begawan (Brunei Darussalam) and lowest in Metro Manila (Philippines). The differences in vehicle per capita ratios in ASEAN countries can be explained by several factors, including the relative affluence of the people, the

Table 2.4. Number of motor vehicles in selected ASEAN capital cities, 1993. (Source: ASEAN, 1997.)

ASEAN capital cities	Population	Number of motor vehicles	Vehicles per 1000 population
Bandar Seri Begawan	276,300	153,351	555
Jakarta	9,200,000	1,072,500	117
Kuala Lumpur	2,000,000	733,083	370
Metro Manila	8,900,000	675,000	75
Singapore	2,873,800	584,000	203
Bangkok	7, 100,000	2,500,000	350

adequacy and efficiency of public transport systems, the pace and extent of the development of suburban settlements, and specific government policies.

7 Effects of air pollution

The effects of air pollution can be classified into several types, with the three most important being damage to plant and animal life, damage to human health and ecosystem disruption. This last effect has captured world attention because of the impacts on local and regional climate and the potential danger of greenhouse gases. The damaging effects of air pollutants on habitat and forests have been of great concern especially in view of what has been learnt about the effects on European and North American forests.

Air pollution can affect humans in a number of ways, and thus represents a health hazard. Because of this hazard, much attention has been given to human-related health problems. Research has provided overwhelming evidence that high levels of some air pollutants can be fatal to the very young, the old and those already weakened by heart and lung disease. The two significant health impacts of air pollution are premature mortality, largely from exposure to high levels of fine particulate matter, and excess cases of chronic bronchitis and other respiratory disorders. Such impacts have been determined through epidemiological research begun in industrial countries in the 1950s, but more work is increasingly being carried out in developing countries.

A worldwide review of 126 cities in which the levels of particulates exceed WHO guidelines estimated that 130,000 premature deaths and 50–70 million incidents of respiratory illness occur each year due to air pollution (Maddison, 1997). In East Asia alone, there are more than 10,000 deaths in Beijing, and 3000–6000 a year in Jakarta, Seoul and Manila. In monetary terms, these costs total 28% of the urban gross domestic product in Beijing, 8–30% in other Chinese cities, 7% in Manila and Bangkok and 4% in Seoul. Comparable results have been reported in India (40,000 deaths), Cairo (5000 deaths) and Mexico

City (6400 deaths). Therefore, urban pollution is not only a major health concern but also has significant economic implications.

In a study of the effects of suspended particulate matter on daily death rates in India, it was found that while an increase in total suspended particulates (TSP) of 100 $\mu g\ m^{-3}$ increased non-trauma deaths by 6.7% in the USA, the same change increased deaths by only 2.3% in Delhi (Schwartz and Dockery, 1992). The difference can be explained by the role of other causes of mortality in Delhi, making the impact of air pollution less apparent. This is an important study, as it illustrates a fundamental difference between developed and developing countries. In developed countries, many potentially fatal diseases have been eliminated or at least severely curtailed. As a result, the relative importance of other causes increases. In developing countries, such improvements have not been made, so the importance of air pollution as a cause of mortality appears to be less.

8 Relationship between haze episodes and respiratory diseases in Malaysia

The most important pollutant of urban areas in South-east Asia is suspended particulate matter, and the main health complaints associated with the resulting haze are conjunctivitis, bronchitis and asthma. During the haze episodes of 1990, 1991 and 1994 in Malaysia, the incidence of such complaints increased (Ambu *et al.*, 1999).

A study carried out during the 1994 haze episode in the Klang Valley showed a significant correlation between the incidence of asthma attacks and PM_{10} (suspended particulate matter with a diameter of 10 μm or less) concentrations during the second half of September. No correlation was found with other variables, including NO_2. This suggests that during the haze episode when the Air Quality Index was hazardous (that is, when the PM_{10} concentrations were high because of the stable air), the environment became a dominant factor in triggering asthma attacks. In comparison with the beginning of the month, there was an increased incidence of such attacks, and the attacks experienced by affected patients were more severe. There was also a significant relationship between the attacks and the combination of NO_2 and PM_{10} throughout the study period, clearly indicating that environmental pollutants have an influence on the health of asthma patients (JICA, 1993).

A study by the Ministry of Health (Malaysia) of the 1994 episode showed that in the Kuala Lumpur Hospital there was a significant increase in the mean average number of asthma cases for children and adults during the haze period compared with the pre-haze period. Other health indices that are affected by the PM_{10} particles include emergency visits to hospitals for asthma, the use of bronchodilators for asthma by children, respiratory symptoms in children and hospital admissions for pneumonia in elderly people.

The increase in suspended particulate matter concentrations has been attributed to forest fires and motor vehicle emissions in Peninsular Malaysia. This is the main cause of the high concentrations of particulates observed in urban centres. During the haze episode of September 1997, the mean level of PM_{10} in the Klang Valley was $247 \pm 88.7\ \mu g\ m^{-3}$ with a range of 83.08–421.38 $\mu g\ m^{-3}$ (Ambu *et al.*, 1999). However, other pollutants (sulphur dioxide, nitrogen dioxide, ozone and carbon monoxide) did not exceed the critical levels.

The economic impact of damage caused by the haze arising from the 1997 forest fires has been assessed by the Economic and Environmental Programme for South-East Asia (EEPSEA). The report found that the total cost of damage amounted to US$1.396 billion, mainly because of a sharp fall in tourism (EEPSEA, 1998). Indonesia suffered the most damage, estimated at US$1.012 billion, followed by Malaysia and Singapore. These estimates are conservative 'lower bound' figures as some of the damage cannot be placed in economic terms, especially that with intangible values. The detailed methods used to obtain the above estimates and some proposed policy recommendations concerning forest fire management in the region have been described by Glover and Jessup (1999).

9 Conclusions

The state of air quality varies from country to country and is partly influenced by inherent physical factors, policy choices, climate and a country's level of development. Many developing countries have shown a progressive degradation in air quality over the last decade. Among the major contributing factors are rapid urbanization, increased industrialization and rising use of energy, mostly derived from fossil fuels. While these programmes are essential for the economic growth of countries and the improvement of their populations' welfare, they in turn lead to pressures on environmental quality and human health. For some countries, besides confronting the unfinished agenda of addressing traditional environmental health problems – clean water, sanitation and nutrition – air pollution is emerging as the next major environmental problem that needs to be solved. This is especially true for many cities in Asia and Latin America where the urban populations have risen to an intolerable level and the provision of basic services has been inadequate.

The three most important pollutants affecting human health in developing countries are lead, particulate matter and soot. For example, high levels of suspended particulate matter have caused premature deaths and millions of people are affected by pollution-related respiratory illnesses each year. Ultimately, the cost of the impacts associated with air pollution in developing countries may be tremendous in terms of lost productivity and other damage. Therefore, because of health, environmental and economic reasons, the formulation of pollution management strategies in developing countries is critical in order to achieve the goal of sustainable development.

References

ADB (1994) *The Environment Program: Past, Present, and Future.* Asian Development Bank, Manila.

Ambu, S., Sahani, M. and Lim, L.H. (1999) *Impacts of Climate Change on Public Health.* GEF Project on Impacts of Climate Change on Key Economic Sectors in Malaysia. Special Report, Institute of Strategic and Policy Studies (ISIS), Kuala Lumpur.

ASEAN (1997) *First ASEAN State of the Environment Report.* ASEAN Secretariat, Jakarta, Indonesia.

Ayers, G., Gillett, R.W., Selleck, P.W., Marshall, J.C., Granek, H., Leong, C.P., Lim, S.F., Haryanto, H., Tuti, M. and Parry, D. (1997) Acid deposition and haze in Malaysia and Indonesia: causes, relationships and consequences. In: Haron, A.H., Dahlan, T., Mohd. Puat, D. and Amran, M. (eds) *Transboundary Pollution and the Sustainability of Tropical Forests.* ASEAN Institute of Forest Management, Kuala Lumpur, pp. 29–44.

Brandon, C. (1998) *Cities and Health.* Environmental Matters, Autumn 1998, the World Bank, Washington, DC.

Bull, K. and Fenech, A. (1999) International activities to reduce pollution impacts at the regional scale. In: Innes, J.L. and Oleksyn, J. (eds) *Forest Dynamics in Heavily Polluted Regions.* CAB International, Wallingford, UK, and IUFRO Vienna, pp. 219–230.

Cevalos, D. (1996) *El Reino del Auto.* Tierramerica: Suplemento de Medio Ambiente para America Latina y El Caribe. A Publication of UNEP Regional Office for LAC, Year 2, No. 3 June.

Cheang, B.K., Leong, C.P., Ooi, S.H. and Abdul Malek Tusin (1991) *Haze Episode October 1991.* Information Paper No. 2. Malaysian Meteorological Service, Kuala Lumpur.

Chow, K.K. and Lim, J.T. (1983) Monitoring of suspended particulate in Petaling Jaya. In: *Urbanization and Ecodevelopment with Special Reference to Kuala Lumpur.* Institute of Advanced Studies, University of Malaya, Kuala Lumpur, pp. 178–185.

EEPSEA (1998) *Indonesian Fires and Haze of 1997: the Economic Toll.* Special Report by the Economic and Environmental Programme for South-East Asia (EEPSEA), Singapore, and World Wide Fund for Nature (WWF), Singapore.

EIA (1995) *Integrated Energy Annual: 1993.* Energy Information Agency, US Department of Energy, Washington, DC.

ESCAP (1990) *State of the Environment in Asia and the Pacific.* Economic and Social Commission for Asia and the Pacific, Bangkok.

ESCAP (1994) *Statistical Yearbook for Asia and the Pacific.* Economic and Social Commission for Asia and the Pacific, Bangkok.

FAO (1997) *State of the World's Forests 1997.* Food and Agriculture Organization, Rome.

Glover, D. and Jessup, T. (1999) *Indonesia's Fires and Haze. The Cost of Catastrophe.* ISEAS Publications, Singapore.

Goldammer, J.G. (1997) Overview of fire and smoke management issues and options in tropical vegetation. In: Haron, A.H., Dahlan, T., M. Puat, D. and Amran, M. (eds)

Transboundary Pollution and the Sustainability of Tropical Forests. ASEAN Institute of Forest Management, Kuala Lumpur, Malaysia, pp. 189–218.

Government of India (1993) *Environment Action Programme*. Ministry of Environment and Forest, New Delhi.

Hettelingh, J.D., Chadwick, M., Sverdrup, H. and Zhao, D. (1995) *RAINS-ASIA: An Assessment Model for Acid Rain*. Report from the World Bank sponsored project 'Acid Rain and Emissions Reduction in Asia'. World Bank, Washington, DC.

JICA (1993) *Air Quality Management Study for Kelang Valley Region. Final Report* Vol. 2, *Main Report*. Japan International Cooperation Agency, Tokyo.

Leong, C.P. and Lim, S.F. (1997) Meteorological monitoring network of forest fire and smoke management in Malaysia. In: Haron, H.A., Dahlan, T., M. Puat, D. and Amran, M. (eds) *Transboundary Pollution and the Sustainability of Tropical Forests*. ASEAN Institute of Forest Management, Kuala Lumpur, pp.125–136.

Maddison, D. (1997) *A Meta-Analysis of Air Pollution Epidemiological Studies*. Centre for Social and Economic Research on the Global Environment, University College, London.

Malaysian Meteorological Service (1995) *Report on Air Quality in Malaysia as Monitored by the Malaysian Meteorological Service*. Technical Note No. 55. Malaysian Meteorological Service, Kuala Lumpur.

Miller, G.T. (1985) *Living in the Environment*, 4th Edition. Waldsworth Inc., Belmont, California.

Schwartz, J. and Dockery, D. (1992) Increased mortality in Philadelphia associated with daily air pollution concentrations. *American Review of Respiratory Disease* 145, 600–604.

Shah, J.J. and Nagpal, T. (eds) (1997) *Urban Air Quality Management Strategy in Asia: Metro Manila Report*. Technical Paper No. 380. World Bank, Washington, DC.

Sham Sani (1987) Air pollution and air quality management in Malaysia. In: Sham Sani (ed.) *Urbanization and the Atmospheric Environment in the Low Tropics. Experiences from the Kelang Valley Region Malaysia*. Penerbit Universiti Kebangsaan Malaysia Bangi, Malaysia, pp. 95–111.

Sham Sani (1994) Urbanization and the environment. In: *Malaysian Development Experience: Change and Challenges*. National Institute of Public Administration, INTAN, Kuala Lumpur, pp. 456–484.

Sham Sani, Cheang B.K., Leong, C.P. and Lim, S.F. (1991) *The August 1990 Haze in Malaysia*. Technical Report No. 49. Malaysian Meteorological Service, Kuala Lumpur.

Shrestha, R.M., Bhattacharya, S.C. and Malla, S. (1996) Energy use and sulphur dioxide in Asia. *Journal of Environmental Management* 46, 359–372.

UNEP (1997) *Global Environment Outlook – 1. United Nations Environmental Programme, Global State of the Environment Report 1997*. Oxford University Press, New York.

UNEP/WHO (1992) *Urban Air Pollution in Megacities of the World*. Blackwell, Oxford.

United Nations (1995) *Compendium of Human Settlement Statistics 1995*, 5th Issue. DESIPASD and Habitat, United Nations Publications, New York.

Vancura, K., Raben, G., Gorzelak, A., Mikulowski, M., Caboun, V. and Oleksyn, J. (1999) Impact of air pollution on the forests of central and eastern Europe. In: Innes, J.L. and Oleksyn, J. (eds) *Forest Dynamics in Heavily Polluted Regions*. CAB International, Wallingford, UK, and IUFRO, Vienna, pp. 121–146.

World Bank (1996) *Livable Cities for the 21st Century. Directions in Development.* World Bank, Washington, DC.

World Bank (1997a) *Urban Air Quality Management Strategy in Asia: Metro Manila Report.* Technical Paper No. 380. World Bank, Washington, DC.

World Bank (1997b) *Urban Air Quality Management Strategy in Asia: Greater Mumbai Report.* Technical Paper No. 381. World Bank, Washington, DC.

World Bank (1998) *The World Development Indicators 1998.* World Bank, Washington, DC.

World Resources Institute (1994) *World Resources 1994–1995. A Joint Report by WRI, UNEP and UNDP.* Oxford University Press, New York.

World Resources Institute (1996) *World Resources 1996–1997. A Joint Report by WRI, UNEP, UNDP and the World Bank.* Oxford University Press, New York.

World Resources Institute (1998) *World Resources 1998–1999. Environment Change and Human Health. A Report by World Resources Institute, UNEP and the World Bank.* Oxford University Press, Oxford.

Air Pollution Problems in the Forested Areas of Mexico and Central America

3

L.I. de Bauer,[1] T. Hernandez-Tejeda,[2] and J.M. Skelly[3]

[1]Instituto de Recursos Naturales, Montecillo, Mexico; [2]Instituto Nacional de Investigaciones Forestales, Agricolas y Pecuarias, SAGAR, México DF, Mexico; [3]Department of Plant Pathology, The Pennsylvania State University, University Park, Pennsylvania, USA

Mexico and the Central American countries share many historical and cultural features. Their climatic conditions as well as their vegetation communities are common in some geographical zones, such as the southern Pacific Mexican coast south to Panama. It is interesting and important to point out that there are favourable conditions for the efficient interchange of pollutants among the three North American countries because of the continuous biogeographical area from central Mexico north through the USA to Canada.

Mexico City illustrates the air quality crisis associated with the ever-increasing environmental problems of the major cities of developing countries. Pollution impacts are mainly seen in the form of oxidant-induced damage to the forested areas downwind of the Valley of Mexico, with sensitive trees such as pine species, sacred fir and 'capulín' black cherry having been studied by several investigators. Some relevant information derived from those studies as carried out in the Chichinautzin Range, especially at the Desierto de los Leones National Park, and Ajusco area is presented here. Air pollutant-induced symptoms and observed forest changes closely resemble the well-described phenomena associated with oxidant problems in the San Bernardino Mountains of southern California.

In the long term, it is likely that the effects of urbanization on the forested areas near the megacity will increase, causing a tremendous change in the adjoining natural ecosystems through the disappearance of sensitive species such as *Pinus hartwegii*. A decline in precipitation, together

 Air Pollution and the Forests of Developing and Rapidly Industrializing Regions (eds J.L. Innes and A.H. Haron)

with the resultant water deficiency, wildfires and increased outbreaks of pests such as bark beetles, may contribute to a shift towards sparser forest vegetation and possibly grasslands.

1 Geographical and climatic description of the area

Mexico is located at the confluence of two large biogeographical regions – the neoarctic and neotropical; whereas Central America could be assigned mainly, or as a whole, to the neotropical. Geographically, Mexico belongs to the North American subcontinent and therefore shares land area with two highly developed countries, the USA and Canada. This geographical area is greatly influenced by important phenomena such as the two high pressure systems that climatologically dominate airflow patterns: the Pacific and the Atlantic permanent highs (Husar, 1985). As for the rest of North America, two mountain chains exist in north–south and east–west directions, comprising the mountain barriers of the central plateau, mainly between 1500 and 2000 m in altitude (Bryson and Haze, 1974; Husar, 1985).

According to Husar (1985), in North America east of the Rocky Mountains, air sweeps southward from the frozen Arctic and northward from the warm tropical seas. The plains of the north and east provide an unobstructed path for north–south excursions of both arctic and tropical air masses. On average, cold and dry arctic air reaches mid-Canada in the summer and the northern USA in the winter. Moist and warm tropical air is restricted to Mexico and to the south-eastern USA in the winter, but reaches south-eastern Canada in the summer. Mild and dry Pacific air penetrates eastward between the Arctic and Atlantic air masses.

These meteorological processes have to be considered when regional pollutant dispersion analyses are undertaken. Some pathogens can travel all the way from Mexico to Canada, as described in the early years of the 20th century for the spores of the stem rust fungus (*Puccinia graminis-tritici*) of wheat (Stakman and Christensen, 1946).

Considering the wind patterns and the anthropogenic emissions in the context of the North American subcontinent, it is possible (although data are lacking) that the more northern areas of Mexico may be influenced by pollutants generated in Canada and the USA (Middleton and Haagen-Smit, 1961). Conversely, it is also possible that some pollutant transport towards the north occurs from Mexico. For example, smoke originating from the catastrophic forest fires during spring 1998 in Mexico and other Central American countries was readily transported from the Yucatan Peninsula, Mexico, Guatemala and Honduras to Florida and Texas, USA.

A closer look at wind directions within the country shows that, for the most part, Mexico is under the influence of the highly humid trade winds that traverse the country from the east and north. However, during the winter season, dry winds from the north-east and west are present in the north, west and central sections of Mexico (Husar, 1985). Nevertheless, because of interference by mountains, valleys and weather-related atmospheric depressions, the actual wind direction is quite variable from one area to another and sometimes even between areas immediately adjacent to each other. Therefore, with few exceptions, most parts of Mexico exhibit irregular airflows. Along the coastal areas of the Atlantic and Pacific Oceans, tropical cyclones with trajectories originating over the oceans produce considerable turbulence over large areas from June to October. In the extensive arid areas of the north and north-west sections of Mexico the winds are not unusually strong, but these areas frequently suffer from the impacts of desert storms (Rzedowski, 1978).

In the Valley (or Basin) of Mexico, the predominant wind direction throughout the year comes from the north and north-east (Jáuregui, 1958). This information has been very useful in characterizing the occurrence of photochemically generated air pollutants and the transport of their respective precursors within and throughout the Valley of Mexico.

Below the southern border of Mexico, the countries and geographical area known as Central America have several elements in common with one another and with Mexico, such as their historical background and the geographical conditions. The Central American isthmus includes the strip of land from the Mexican to the Colombian borders with the following countries: Guatemala, Belize, Honduras, El Salvador, Nicaragua, Costa Rica and Panama. Also, Cuba, La Española, Jamaica, Puerto Rico and several other small islands are normally included when considering Central America. Other elements in common for Central America are the mountain chains, volcanism and strong hurricanes, all of which are characteristics that unify this particular geographical region (Oceano, 1993).

Although the Latin American countries share a heritage and cultural and social interchange is very dynamic, there is a need for improved scientific communication among the countries. With few exceptions, it is difficult to obtain information that could be useful in recognizing problems of concern across the whole region. This is particularly true for air pollution.

2 Historical background

The evolution of the pre-Columbian societies in Mexico occurred between 400 BC and AD 900. Some of the most prominent societies were the Olmeca, Teotihuacana, Totonaca, Mixteca and Maya. Between the 9th and 13th centuries, the migrations of the Nahuatl people (a new group) took place. Later on, another group known as the Toltecas unified the Valley of Mexico. As a second movement of the Nahuatl people (the Aztecas or Mexicas), they were able to build a great empire.

The first incursion of the Spaniards occurred between 1512 and 1518. Thereafter, Hernan Cortes conquered the Aztec empire (1519–1521). A second phase of the Spanish conquest took place between 1529 and 1545, and it was in this period (1535) that the New Spain Viceroyalty was established. During the second half of the 16th and during the 17th and 18th centuries, colonialism became established and the North American territory was further explored.

In 1810, the Miguel Hidalgo insurrection took place and in 1821 the independence of Mexico from Spain was recognized. From 1824 to 1850, civil wars occurred and in 1848 the Guadalupe Hidalgo Treaty dealt with Mexico's loss of the war against the USA. As a result of this treaty, the northern territories of Mexico were ceded to the USA.

The civil war known as 'Reforma' gave power to Benito Juarez from 1857 to 1860. Between 1861 and 1867, a French invasion imposed Maximilian I as emperor, but he was executed on Juarez's orders. From 1877 to 1911, the Porfirio Diaz dictatorship took place; the Mexican Revolution against Diaz occurred from 1911 to 1920. From 1924 to 1940, Elias Calles and Cardenas gave continuity to some of the revolutionary principles (Oceano, 1993). The former founded the official party that has ruled the country more than 70 years.

From being predominantly rural at the turn of the century, Mexico has changed into a semi-industrialized country. A very important factor in this change has been its rapid population increase following the 10-year revolution, i.e. during the last 80 years. The population of Mexico in 1920 was 20 million inhabitants; today it is close to 100 million. This increase in population has had remarkable consequences in terms of increased demands for energy, natural resources, transportation, urban expansion, recreation and especially for increased food production. In relation to forest ecosystems, these issues mean that substantial areas of forest are being converted into agricultural lands and large, expansive urban areas. This phenomenon is expected to continue for several years (Calderón and Hernández, 1987; Pick and Butler, 1997).

In 1992, free trade agreements were signed between the USA, Canada and Mexico; these important agreements have opened a new era of cooperation and opportunities for the three countries. However, as a logical consequence of the policies involved, a number of new industrial activities that are currently common in the two more developed countries will in future contribute to increased pollution problems in Mexico.

In the rest of Central America, progress has also been sporadic. Although the isthmus became a cultural and political entity during the classic period of the Mayas, the Central American unit was dissolved very soon after the Spanish conquest, in a drastic manner, by means of three different forms of colonization: the first from Haiti, the second from Mexico and the last from Panama. Since then, there have been frequent disagreements between those advocating small, independent countries and those seeking to unify the Central America region.

Some examples illustrate the Central America political situation. After the disintegration of the General Captaincy of Guatemala (which during that time included the colonies of Guatemala, Honduras, El Salvador, Nicaragua and Costa Rica), the attempt of the United Provinces of Central America (1823–1838) to create a Major Republic of Central America (integrated with Honduras, Nicaragua and El Salvador) failed. In 1898, the old provinces were attacked by several countries (e.g. the North American invasion led by William Walker in 1855, who in 1857 attempted to take over the isthmus). As a consequence, these events converted Central America into a series of small republics, victims of domestic oligarchies and foreign-owned fruit enterprises.

During the second half of the 20th century, social and political upheavals have been the dominant problems in Central America and in the Caribbean countries. Meanwhile, in some countries, revolutionary movements have taken place (Cuba in 1958 and Nicaragua between 1979 and 1990). Only Costa Rica, since 1949, has had a democratic and peaceful development (Oceano, 1993).

3 Forest ecosystems: present and related environmental issues

The forests of Mexico are the result of millions of years of evolution, and are a testimony to the perseverance and persistence of the plant life on our planet. Their origins go back 400 million years when the first plants emerged from the seas and dominated the continents. This was an advance in evolution comparable only to the complex diversification that occurred with the first appearance of flowering plants 150 million years ago (SARH, 1994a). To explain the rich mixture of today's Mexican flora, it must be noted that for millions of years this geographical region was a bridge for many 'migrations' of plant life from North American to South American regions and vice versa.

Modern Mexico is the home of some 30,000 of the estimated 300,000 plant species around the world. This is the result of the combination of several factors such as the geographical location, mountainous topography, climate and soils. The great variety of flora was fully valued by ancient cultures, as recorded in several prehispanic codices that testify to the close harmony between meso-American people and their environment (SARH, 1994b).

The Mexican and Central American woodlands (Fig. 3.1) have been classified (SARH, 1994a,b) as:

- cold and temperate zone forests
    - conifer and oak forest
    - montane rainforest

Fig. 3.1. Distribution of the Mexican and Central American forest communities. The Valley of Mexico is located in the central part of the country, which includes the Federal District and part of the states of México, Hidalgo, Tlaxcala and Puebla. The lowest part, a lacustrine plain, has an average elevation of 2240 m above sea level. To the east, west and south, a succession of elevated volcanic ranges surround the basin. The highest peaks, Popocatepetl and Ixtaccihuatl, with altitudes of 5465 and 5230 m, respectively, are located to the east.

- tropical forests
  - tropical evergreen forest
  - tropical deciduous forest
- thorn forests
- mangrove swamp.

The Valley of Mexico is located in the central part of the country, which includes the Federal District and parts of the states of Mexico, Hidalgo, Tlaxcala and Puebla. The lowest part, a lacustrine plain, has an average elevation of 2240 m above sea level. To the east, west and south, a succession of elevated volcanic ranges surround the basin. The highest peaks, Popocatepetl and Ixtaccihuatl, with altitudes of 5465 and 5230 m, respectively, are located to the east.

According to prehispanic records, the major environmental zones that existed in the basin were (Millon, 1970): lake system (important for migratory birds); saline lake shore (characterized by halophyllous plants); deep-soil alluvium (covered by sedges and swamp cypresses); thin-soil alluvium (dominated by grasses and agaves); upland alluvium (occupied by oaks and acacias); lower, middle and upper piedmont (mainly oak forests); and sierras (sites above 2700 m with conifer forests). Few of those original ecosystems remain. Among them, the sierras occupy sites above 2700 m and harbour temperate plant communities very rich in pine species, firs and junipers. These communities are surviving, but they are being adversely affected by the uncontrolled cutting of trees and by their exposure to high concentrations of air pollutants. In addition, in all directions, but more severely to the south and west, the expanding population of Mexico City is invading the slopes of the surrounding mountains, eliminating all vegetation – including, in certain parts, the conifer forest cover (Calderón and Hernández, 1987; Ezcurra and Mazari-Hiriart, 1996). This is a worrying development and the disasters (major loss of life following massive mudslides and debris flows) that affected Venezuela in late 1999 could be repeated in Mexico.

3.1 Cold and temperate zone forests

The cold and temperate zone forests of Mexico are located in the highest parts of the country, on the main mountain ranges of the Sierra Madre Occidental, Sierra Madre Oriental, Eje Neovolcanico, Sierra Madre del Sur and Sierra Madre de Chiapas. The great height of these mountain chains, the abundant rainfall and the air currents that cross them all help to determine the level of the humidity and the typical features of the indigenous forest vegetation (SARH, 1994a,b).

Conifer and oak forest

This type of forest is characterized by a wealth of forms, a wide range of species, numerous microclimates and a rich species diversity on the forest floor. Conifer species and oak species grow in cold, temperate and semi-humid conditions. Coniferous forests are composed mainly of different species of pine (*Pinus* spp.) and fir (*Abies* spp.). Pines are the most common conifers in the forest because of their ability to vary and adapt. They grow at altitudes of between 1500 and 5000 m and are widely distributed over the country, except for Tabasco and the Yucatan Peninsula. Forty-two species, 22 varieties and nine sub-types of pines have been recorded in Mexico (SARH, 1994a; Hall *et al.*, 1996). There are also eight species and four varieties of fir in Mexico (Martínez, 1963).

Most of the oak forests grow at between 1200 and 2800 m and are found in all the states, except for the Yucatan Peninsula. There are over 150 oak species and varieties, which ranks Mexico foremost in the world for the diversity of these species, most of them belonging to the genus *Quercus*. The total surface area covered with this type of forest is 30.4 Mha (SARH, 1994a,b; Hall *et al.*, 1996). Oak forests can also be found in the mountains of Central America (Burkart *et al.*, 1995).

Montane rainforest

This forest type is one of the most beautiful ecosystems in existence. The diversity of plant and animal life is immense, while the dense vegetation and the almost permanent mist that shrouds the forests during parts of the year give it the appearance of an enchanted forest. The distribution of this forest type depends more on climate than on soil type; the forest is dense, with trees averaging 15–35 m tall, with some as tall as 60 m. The distribution of this forest type is limited and fragmentary, stretching from south-east Tamaulipas to north Oaxaca and Chiapas. On the Pacific slopes, it is even more widely scattered, from the north of Sinaloa to the Valley of Mexico. The main plant species are tree ferns (e.g. *Cyathea mexicana*), magnolias (*Magnolia* spp.), liquidambar (*Liquidambar* spp.) and handflower trees (*Chiranthodendron pentadactylon*). Another important characteristic feature is the abundance of epiphytes (primarily bromeliads and orchids). The total surface area covered with this forest type is 1.4 Mha (SARH, 1994a,b; Hall *et al.*, 1996).

3.2 Tropical forests

Mexico's tropical forests are located on the mountain slopes and coastal plains of the Pacific Ocean and the Gulf of Mexico. Climatic differences on each side of the country determine the type of tropical forest. On the Gulf Coast, heavy rainfall and high humidity produce a vegetation cover different from that produced in the strongly seasonal areas of the Pacific coast (SARH, 1994a,b).

Tropical evergreen forest

Tropical evergreen forests grow in hot, humid regions where there is high water availability all year with practically no seasonal variation. This type of forest grows at altitudes ranging from sea level to 1000 m and occasionally up to 1500 m. The tropical evergreen forest area in Mexico includes the Huasteca region in south-east San Luis Potosi, northern Hidalgo and Veracruz, and extends as far as south Campeche and Quintana Roo, taking in parts of

Oaxaca, Chiapas and Tabasco. The forest is rich in vines and epiphytic plants. The vigour of the plant life is exemplified by trees 30–45 m tall. The total surface area covered with this forest type is 5.8 Mha, which includes tropical sub-deciduous forests (SARH, 1994a,b; Hall *et al.*, 1996). This type of vegetation can be found along the Pacific and Atlantic Ocean shores of Mexico and Central America (Burkart *et al.*, 1995).

Tropical deciduous forest

This is a forest of seasonal contrasts: it is transformed from a grey, leafless, apparently dead wilderness in the dry season from November to March or April to lush vegetative growth during the rainy season from May to September or October. This forest is typical of hot climates with very dry seasons that last between 5 and 7 months, followed by a short rainy season. The tropical deciduous forest grows at altitudes ranging from sea level to 1900 m. On the Pacific coast, it stretches from south Sonora and south-east Chihuahua to Chiapas, including part of Baja California Sur. There are also three isolated belts of this vegetation in the states of Tamaulipas, San Luis Potosi and Veracruz. The forest is relatively low and the trees often have crowns of a diameter that equals or exceeds their height, which varies between 5 and 15 m.

Some of the most characteristic trees of such forests are gourd trees (*Acacia* spp.), and others known as 'guamuchiles' (also known as Madras thorn, Manila tamarind) (*Pithecellobium dulce*). Some species are interesting because they have become economically important, such as the sisal hemp (*Agave fourcroides*, mainly on the Yucatan Peninsula), 'bursera' (*Bursera* spp., an aromatic wood used for handicrafts) and 'cuachalalate' (*Amphipterygium glaucum*, used for the medical qualities of its bark and roots). The total surface area with this type of forest is 10.9 Mha.

There are some other types of tropical forest of lesser importance in the country; these are variously termed 'palmar', 'sabana', 'selva de galeria' and 'selva fragmentada', and cover a surface area of 7.6 Mha (SARH, 1994a,b; Hall *et al.*, 1996). These types of vegetation are also present along the Pacific Ocean shore, from Mexico to Central America (Burkart *et al.*, 1995).

3.3 Thorn forests

The thorn forest is capable of existence in extremely arid conditions. The trees stand in sharp contrast to the parched environment. The forest type is mainly composed of spiny trees and scrub vegetation usually found in the driest and hottest land areas and in some arid lands. Associated species occupy a belt on the southern edges of the Mexican high plateau and across some of the arid regions of the country, such as the coastal plain of Sonora and Sinaloa. This forest type is also found in the Huasteca and Bajio regions, and extends from the north to the south of the country, reaching Yucatan. Numerous thorny species grow between 4 and 15 m tall, and the arboreal stratum is often dense. Various types of trees and other large plants can be found here, including mesquite (*Prosopis* spp.), yucca (*Yucca* spp.), buckthorns (e.g. Mexican crucillo, *Condalia warnockii*; and greythorn, *Ziziphus obtusifolia*), cow's tongue (*Cercidium* spp.), cacti and several others typical of this ecosystem; this type of forest covers a surface of 17.2 Mha (SARH, 1994b). This vegetation is also present in the dry areas of the Pacific Ocean shore of Central America (Burkart *et al.*, 1995).

There is a xerophyllous thorn forest in the central part of Mexico, which covers a very large area (41.3 Mha), but the species present are mainly shrubs and/or 'chaparrales' (SARH, 1994a,b).

3.4 Mangrove swamps

Mangrove swamp forests comprise a peculiar type of forest where two diametrically opposite but intimately related natural environments meet – the sea and *terra firma*. Mangrove swamps are widely distributed along the coasts of the world's hotter regions; the forests are dominated by trees from 2 to 25 m tall. Mangrove swamps are found mainly along the shores and the coastal lagoons, in sheltered bays and at the mouths of some rivers where there is an intermixing flow of fresh and salt water. The swamps are scattered along the Pacific coast from halfway along the Baja California Peninsula to Chiapas. On the Gulf and the Caribbean coasts, these forests are more continuous, stretching from Tamaulipas to Quintana Roo. The most extensive mangrove swamps are on the coastal plain of Nayarit, where they are known as the National Marshes. Although there are approximately 70 species of plants and trees in the world that typically grow in mangrove swamps, there are only four in Mexico: red, black, white and button mangroves (*Rhizophora mangle, Avicennia germinans, Laguncularia racemosa* and *Conocarpus erecta*). Despite their high ecological value and enormous economic potential, for a long time the swamps were considered as idle lands or, worse still, unhealthy. This type of forest covers an area of 0.7 Mha (SARH, 1994a,b); vegetation of this type can be found along the Pacific Ocean shore from Jalisco and Nayarit all the way to Panama (Burkart *et al.*, 1995).

3.5 Other forested areas in Mexico

Finally, it is important to mention that there are some types of vegetation growing in fresh waters and still others in salty soils, covering a total surface of 4.2 ha. There is also a very important area of disturbed forests comprising an area of 22.2 Mha. The total area of forested ecosystems in Mexico, according to the six main types of forests and several others considered here, is 141.7 Mha, comprising 72.05% of the total land area of 196.7 Mha (SARH, 1994b).

The extraordinary diversity of the forested areas of Mexico is declining; loss of diversity is especially found in places where the local population depends on wood as a fuel. A similar situation is present in Central America.

4 Regional impacts of air pollution

As already stated, the environment of the Valley of Mexico where Mexico City is located has undergone many important changes during recent decades. This extensive urban and industrialized area suffers from very high air pollutant exposures.

4.1 Signs of environmental deterioration

The air quality of Mexico City has become very poor during the last 30 years and its pristine atmosphere is unfortunately now just a legend. According to Jáuregui (1983), while there was a clear deterioration in visibility between 1940 and 1974, there seems to have been a gradual improvement since 1975 as a result of government policies controlling specific industrial particulate emissions. The trend is evident from a decrease in the frequency of days with visibility in the 0.5–1.0 km range. There are exceptions to this trend: in 1998, visibility was severely impaired by smoke originating from forest fires brought about by severe drought conditions. Fine particulate aerosols are the main cause of reduced visibility. Despite government efforts, the abundance of suspended particulate matter poses very severe problems for the Valley of Mexico (Quadri and Sánchez, 1994).

Several natural and anthropogenic characteristics have been blamed for the environmental deterioration in the Valley of Mexico. All of them induce certain consequences (Table 3.1). Most of these factors interact in such a manner as to make the Valley of Mexico an extremely efficient large-scale site for the generation of oxidant-type precursor pollutants and photochemically generated air pollutants such as ozone.

Previously, Bauer (1972) investigated the responses of sensitive indicator plants and reported that photochemical oxidants were the most important phytotoxic air pollutants within the urban complex. This remains the case. The metropolitan monitoring network indicates that ozone is the most prominent air pollutant, although particulate matter is also a very important component of atmospheric pollution. The current air quality situation has led to comparisons between the air quality of Los Angeles, California (USA) and

Table 3.1. Natural and anthropogenic characteristics of the Valley of Mexico.

Characteristic	Consequence
Natural	
Flat basin surrounded by mountains	Restricted air circulation
Considerable altitude (2250 m asl)	High UV-B levels
High frequency of light winds (< 1.6 m s^{-1})	Poor dispersion
Anthropogenic	
Human concentration	Depletion of natural resources
Presence of a number of diverse industries (over 40% of the country)	Emissions (point sources)
Proliferation of motor vehicles (approx. 5,000,000)	Emissions (mobile sources)

Mexico City. This led Miller *et al.* (1994) to compare seasonal ozone concentrations between the Mexico City and Los Angeles regions. They found that ozone concentrations were higher in the San Bernardino Mountains during the summer oxidant season than in the Desierto de los Leones National Park. However, there was a respite from exposure during the winter season in the San Bernardino Mountains; this respite does not take place in the highly impacted Desierto de los Leones area. As estimated for 1997 by Quevedo *et al.* (1999), the annual average daytime (0800–1900 h) value was 0.252 ppm (252 ppb) with peaks up to 0.3029 ppm (303 ppb) for the monitoring site within the mountains of the Desierto de los Leones National Park.

According to Quadri and Sánchez (1994), more than 85% of the registered vehicles in the Federal District (DF) are private. These data indicate the importance of this means of transportation, although they are used by only 16% (4,400,000 per person per day) of the daily commuters, whereas with just 7% of the registered vehicles, the public transport (including the subway) transports 84% of the commuters. It is also important to point out that 45% of the private cars are 10 years or more old and the traffic in Mexico City is slow and congested. Approximately 10 years ago, Romero (1987) assessed an average speed of <12 km h^{-1} for vehicles in Mexico City.

Mobile sources are considered to be the main contributors to the precursors of photochemical smog, and are estimated to contribute 60% of the total atmospheric pollutant load in Mexico City (Velasco Levy, 1970; Quadri and Sánchez, 1994). Wirth (1997) pointed out that the privatization of transport in Mexico City exacerbated traffic congestion and air pollution.

Bravo *et al.* (1987) and Bravo and Torres (1995) assessed the most important pollutants emitted from the metropolitan zone of Mexico City to be hydrocarbons (525,000 t $year^{-1}$), sulphur dioxide (411,000 t $year^{-1}$) and nitrogen oxides (132,000 t $year^{-1}$). The introduction in 1991 of catalytic convertors (DDF and Limusa, 1997), which are now used in about half of the vehicles, has probably resulted in improved air quality, but the relevant data are difficult to obtain.

Until 1987, lead was considered to be the most dangerous pollutant; after 1987, the national oil company, Petroleos Mexicanos (PEMEX), introduced a low-lead fuel. According to the official monitoring network for the metropolitan area, ozone is now the air pollutant of greatest concern. Therefore, daily monitoring is based on measurements of this pollutant.

Within the Valley of Mexico, the early work of Jáuregui (1958) demonstrated that the prevailing winds blow from north-north-east; high ozone concentrations therefore mostly affect the southern residential areas of Mexico City, as well as the forests located to the south of the metropolitan area. The highest ozone values recorded by the Mexico City monitoring network are from the monitoring station located at El Pedregal. In this part of the city, peak ozone concentrations are recorded between 1200 and 1400 h; during 1994, peak hourly values exceeded the maximum allowable standard (0.11 ppm) on 345 days, according to the official report of the Mexican National Institute of Ecology (INE, 1995).

The air quality problems of Mexico City are of considerable concern for human health. In the southern areas, photochemical oxidants pose a real hazard as a result of the frequent concentrations above permitted limits (INE, 1995). Rosas *et al.* (1995) found that there is a gradient in atmospheric lead from north to south, decreasing from about 1.2 to 0.5 $\mu g\ m^{-3}$ toward the southern suburbs. A similar situation exists for particulate matter emissions and deposition (Figs 3.2 and 3.3).

4.2 Natural sources: volcanoes

At present, Popocatepetl (19.02°N, 98.62°W) is the only active volcano close to the Valley of Mexico. High concentrations of sulphur dioxide (SO_2) were most recently recorded on 1 February 1994. High levels of SO_2 emissions are known to indicate a reactivation of volcanic activity, with a further associated risk of magmatic activity (Galindo *et al.*, 1995).

On 21 December 1994, after several decades of relative inactivity, the Popocatepetl volcano began a new eruption. After the eruption, there were continuing seismic conditions with plume emissions and SO_2 flux. Between February 1994 and January 1995, an SO_2 flux measurement was made on the plume of Popocatepetl with results showing that the maximum gas (SO_2) and particles (ash) emissions occurred 72 h after the onset of the eruption (24 December 1994). These values were an average of 3961 t SO_2 day^{-1} (recorded during the fly ash emissions between 1238 and 1344 h) and a maximum value of 4555 t SO_2 day^{-1}.

The maximum rate of emission was closely associated with the maximum tremor amplitude and the maximum number of seismic type B events. On 5

Fig. 3.2. The smog in the Mexico City Basin often forms a very sharp boundary, as shown here.

Fig. 3.3. Very poor air quality is now a characteristic of Mexico City.

November of the same year, the average emission of SO_2 was 1261 t day^{-1}. The SO_2 baseline for the 1994 period was about 1000 t SO_2 day^{-1} (Galindo *et al.*, 1995). There is very little information about the emissions of SO_2 from active volcanoes in other areas of Mexico or for other parts of Central America.

5 Detected impacts

The phenomenon of urban agglomeration common all over the world, but exacerbated in developing and non-industrialized countries, finds one of its best examples in Mexico City, one of the largest cities of the world. The uncontrolled demographic and urban growth is having tremendous ecological consequences in this once beautiful area. When considering the environment as the sum of all physical, biological and social factors for a given area, Mexico City can be judged from all three points of view, but its demographic explosion can probably be taken as its greatest challenge.

Ezcurra and Mazari-Hiriart (1996) reported that Mexico City had a population of 18 million people in 1995. Given current trends in population growth and the increased rural to urban migration created by the economic crisis of the late 1990s, the population of Mexico City will soon be above 20 million. The ever-increasing demands for services from the population, in terms of housing, electricity, water and food supplies, health care and education, are enormous and are driving the local and federal authorities towards a critical situation. The ecological consequences of approximately 20 million people occupying a limited space consisting of roughly 12% of the Mexico City basin (7500 km^2) are linked to the high risk of atmospheric degradation, exhaustion of the water supply and a collapse of urban services (Fig. 3.4).

The city and federal authorities are well aware of the very important problems affecting the megalopolis. There are several regulations aimed at reducing urban atmospheric deterioration, such as the control of industries and vehicle emissions. The 'once a week does not circulate' regulation established for older cars without catalytic convertors illustrates governmental efforts to mitigate the problem.

5.1 Responses of forest ecosystems and vegetation located in the southern mountainous areas to air pollution generated in Mexico City

Meteorology, topography and the concentration of the human population interact with some other important factors in the Valley of Mexico to produce high concentrations of air pollutants such as ozone, nitrogen oxides and peroxyacetyl nitrate as well as particulate matter (Bauer and Krupa, 1990; Table 3.1). It was noted as early as 1971 that sensitive plants, after being exposed at different places in the metropolitan area, would show symptoms of air pollution injury, such as bronzing of the underside of leaves of romaine lettuce (*Lactuca sativa* cv. Romaine) and O_3-characteristic leaf-stippling on tobacco Bel W-3 (Bauer, 1972). These results indicated the very probable presence of phytotoxic oxidants and their primary pollutant precursors. While determining where the urban plume was deposited, Krupa and Bauer (1976) found and described symptoms of photochemical oxidant injury on two pine

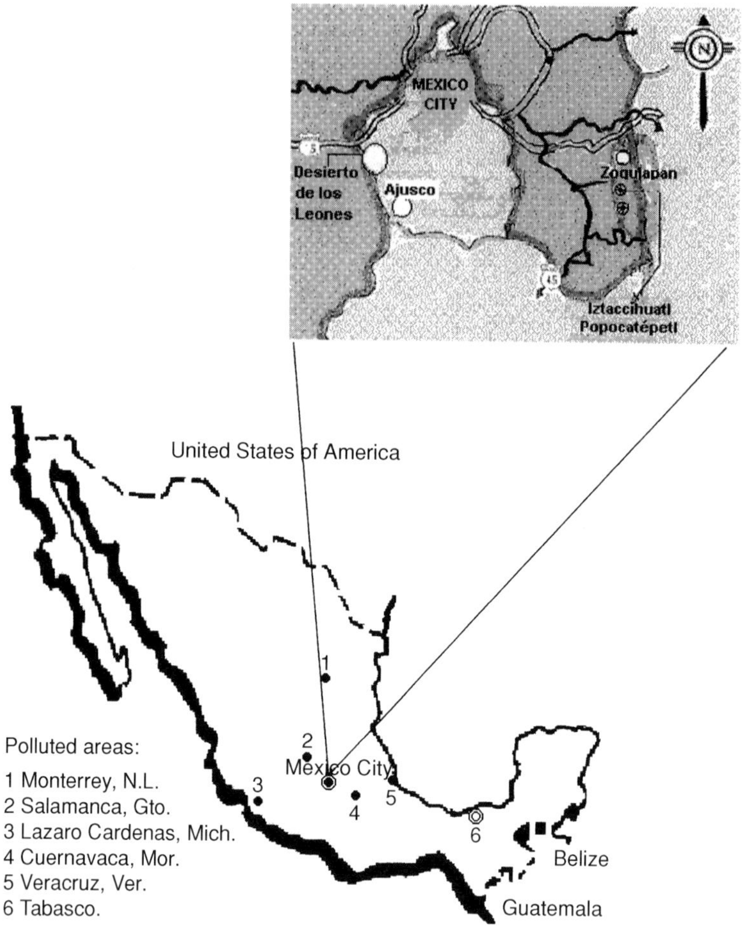

Fig. 3.4. Mexico Valley and location of five-polluted cities, one state (Tabasco) and three National Parks: Desierto de los Leones, Ajusco and Zoquiapan. The locations of the volcanoes Popocatepetl and Ixtaccihuatl are also indicated.

species, *Pinus hartwegii* Lindl. and *P. leiophylla* Schl. et Cham., at Ajusco, DF, a southern forested area of Mexico City.

The recognition of oxidant-type damage on pine trees was the starting point for a series of research projects carried out over several years, with most studies being located at Ajusco. However, a massive decline of sacred fir (*Abies religiosa* H.B.K. Schl. et Cham.), first observed in 1982 at the Desierto de los Leones National Park, caused alarm because of its magnitude (Fig. 3.5). The park has an area of 1529 ha. It is located to the south-west of Mexico Valley at an elevation of between 2800 and 3800 m (Fig. 3.4). The fir forest is prevalent at an altitude of 2800–3200 m. Precipitation is usually 1300 mm year^{-1} and

Fig. 3.5. Severely impacted forests of *Abies religiosa* in the Desierto de los Leones, close to Mexico City.

mean temperature is between 7 and 15°C. The soils of the park are generally rich, with a pH of 5–7 (Vazquez, 1986).

During July and August 1980, observations were made in the most critically polluted areas of Mexico, with the help of Dr Sagar V. Krupa from the University of Minnesota, USA (Fig. 3.4). Some plant species at Ajusco (DF), Xochlmilco (DF), Chapultepec (DF), Lazaro Cardenas (Michoacán), Salamanca (Guanajuato), Cuernavaca (Morelos), Monterrey–Chipinque (Nuevo Leon) and in the states of Veracruz and Tabasco, showed symptoms typical of ozone-induced injury. The most important and frequent foliar injury was related to ozone but in some cases, proximal to oil refineries, injury was ascribed to sulphur dioxide (Bauer and Hernández, 1986).

One case observed in 1980, involving clear symptoms typical of ozone-induced foliar injury, was the damage to *Eucalyptus globulus* Labill. seedlings in newly forested urban areas. To establish the possible cause–effect relationship, seedlings of *E. globulus* were exposed for just 1 day to 0.40 ppm ozone for 2 and 4 h under environmentally controlled conditions. The plants in both treatments developed symptoms consisting of flecking and bleaching on the upper surface of the intermediate-age leaves, and the damage intensity was directly related to the duration of the exposure. The symptoms resembled those observed under natural conditions in southern Mexico City (Hernández *et al.*, 1981). More recently, Martínez and Chacalo (1994) have observed pollutant-induced injury on different tree species within the urban area of Mexico City. However, for these species, further research is needed to establish the cause–effect relationships.

Based on the earlier observations of oxidant-induced injury on pine species, an *in situ* assessment was made of *Pinus hartwegii*, *P. montezumae* var. *lindleyi* Loud. and oats (*Avena sativa* L.) at Ajusco. *P. hartwegii* showed a chlorotic mottling and banding on current-year needles, and senescence and premature shedding of previous-year and older needles. *P. montezumae* had almost the same symptoms, but they were less severe. In both species of pine, the most severe injury was apparent at the end of spring and beginning of summer. The oxidant-induced injury to oat leaves took the form of chlorosis and single surface and bifacial necrotic bleaching (Hernández, 1981). In addition, transverse leaf sections from *P. hartwegii* revealed histological changes, with collapse and necrosis of the vein tissue. In oats, the injury involved plasmolysis of the mesophyll cells and collapse of the cell walls. Cells adjacent to the stomata also showed severe plasmolysis (Hernández, 1981).

Studies of *P. hartwegii* have shown that this species, found at 2850–3500 m, is highly sensitive to ambient ozone exposures. It is one of the nine native pine species in the Valley of Mexico and, apparently, it is the most sensitive to O_3 (Hernández and Bauer, 1984). Ozone injury has also been noted on many herbaceous plants at Ajusco and along the highway traversing south to Cuernavaca (Morelos) (Hernández, 1984).

In an attempt to trace the trajectory and impacts of the urban plume, *P. hartwegii* and *P. montezumae* var. *lindleyi* were observed for 2 years at several sites along the mountain road from Mexico City to Cuernavaca. A gradient in symptoms of needle injury was clearly observed, with injury decreasing with increasing distance to approximately 56 km south of Mexico City (Hernández and Bauer, 1984).

In the Ajusco mountains, a second assessment was made to examine the response of *P. hartwegii* and *P. montezumae* to photochemical oxidants. Using a visual scale for the evaluation of oxidant-induced injury to pines (Miller, 1973), foliar injury on *P. hartwegii* was found to be more severe than on *P. montezumae* at the end of 2 years of assessments, and in *P. hartwegii* the injury began 3 months earlier (Hernández and Bauer, 1984).

The response of *P. maximartinezii* to photochemical oxidants has been examined at a tree nursery located in the south-central part of Mexico City. The results showed a homogeneous pattern of health and vigour and only medium sensitivity (based on foliar injury) to photochemical oxidants, although the trees maintained their needles through only a 3-year cycle rather than the normal 5-year period (Hernández and Nieto, 1996).

Chlorophyll was extracted from *P. hartwegii* needles collected at Ajusco; losses in chlorophyll *a* amounted to 34.7% in comparison with those branches protected with charcoal filters. The *b* and *a* + *b* chlorophyll contents showed a smaller loss than the chlorophyll *a* (Hernández *et al.*, 1986). Because of its O_3 sensitivity, as revealed by the extensive chlorotic mottling and banding and loss of mature needles, *P. hartwegii* would be a useful bioindicator for O_3 impacts. This species has also been studied in other places, such as the Desierto de los Leones National Park, where very severe damage can be observed not only in this species but also in sacred fir (Alvarado, 1989; Alvarado *et al.*, 1993). Several investigators working in the Desierto de los Leones National Park have suggested a possible relationship between altitude and the severity of the O_3 symptoms observed on *P. hartwegii* (Bauer *et al.*, 1985; Hernández and Bauer, 1986).

The first symptom seen on needles of declining *A. religiosa* at Desierto de los Leones, possibly induced by ozone, is the formation of small whitish lesions on the upper surfaces of the older needles; later, the lesions turn reddish-brown in colour and the entire needle becomes necrotic. Trees in the sample plots located south and south-east of the Cementerio ravine were the most affected, as indicated by the proportion of dead trees recorded in surveys conducted between 1986 and 1988 (Alvarado, 1989; Alvarado *et al.*, 1993). Histological examination and chemical analysis of damaged and healthy needles of *A. religiosa* have been undertaken. The results indicate a similarity to the palisade tissue damage known to be induced by ozone, as well as phenolic compound accumulations typical of those associated with oxidant injury (Alvarado, 1989; Alvarado *et al.*, 1993). These findings were confirmed by Alvarez (1996). Branches exposed to charcoal-filtered air in chambers did not develop any of the above symptoms. Likewise, needles protected by an anti-transpirant maintained their typical dark green colour and needle retention was longer (Alvarado, 1989; Alvarado *et al.*, 1993).

In additional studies, chemical analysis of the damaged *A. religiosa* needles indicated a deficiency of manganese (average 37 ppm) in relation to healthy needles (average 112 ppm). Soil analysis of rhizosphere samples from declining trees at the Desierto de los Leones has also indicated a deficiency in manganese and zinc in comparison with *A. religiosa* foliage without evident damage (at Zoquiapan) (Alvarado, 1989; Alvarado *et al.*, 1993). However, a later study found no differences in soil manganese or zinc contents between the same places (Castro *et al.*, 1996). Watmough (1997) has argued that some of the lead and cadmium levels from Desierto de los Leones soils are higher (the highest concentration for lead was 247 mg kg^{-1}, and the highest cadmium concentration was 1.34 mg kg^{-1}) than those found in areas within urban Toronto and New York City. In addition, Watmough (1997) noted that the zinc concentrations in tree rings of *A. religiosa* are higher (361 mg kg^{-1}) than within any species cited in the literature.

A major ring-width reduction has been associated with the decline of *A. religiosa* trees, with the decline starting at the beginning of the 1970s (Alvarado, 1989; Alvarado *et al.*, 1993) (Fig. 3.6). Earlier reductions in ring width were also observed during the 1920s and 1930s (Watmough, 1997).

Besides being sensitive to ozone, *P. hartwegii* has demonstrated that the interaction between a biotic pathogen and oxidants can be synergistic. At Ajusco, *P. hartwegii* had the characteristic chlorotic banding and mottling of needles, but also a severe infection by the leaf pathogen *Lophodermium* spp. The

Fig. 3.6. Reduction of ring width (mm) in *A. religiosa*. This symptom is shown by the trees at Desierto de los Leones National Park.

magnitude of the damage was directly related to the age of the leaves and, very clearly, to the exposition of the affected trees towards the urban area (Alvarado and Bauer, 1991).

The severe damage observed on pines first at Ajusco, and later in the Desierto de los Leones National Park located further south-west of Mexico City, resembles the injury to pine trees (*P. jeffreyi* and *P. ponderosa*) in the San Bernardino Mountains of southern California (Miller *et al.*, 1991, 1994; Miller and McBride, 1999). Using a standard procedure for evaluating ozone injury to pines, Miller *et al.* (1996) compared the crown condition of damaged *P. hartwegii* at Desierto de los Leones with the crown condition of a mixture of *P. ponderosa* and *P. jeffreyi* at Barton Flats, California, USA. The ozone injury index revealed that the Mexican pine species had the highest amount of crown injury (upper injury classes) whereas the Californian species were only moderately damaged (lower injury classes). Both situations seem to be similar in their evolution. The patterns of growth of *P. hartwegii* also showed a marked decline in growth since the early 1970s at both sites to the south of Mexico City, i.e. Ajusco and Desierto de los Leones; there were no signs of a recovery (Alarcón *et al.*, 1995).

Broadleaved forest tree species within the urban and surrounding forested areas of the Mexico City airshed also suffer from oxidant-induced stress; symptoms include premature chlorosis and early leaf senescence indicative of season-long oxidant exposures (Bauer and Krupa, 1990). Skelly *et al.* (1997) conducted surveys for foliar injury on capulín black cherry (*Prunus serotina* var. *capuli*) within the Desierto de los Leones National Park. Black cherry is well known as a species that appears to be very sensitive to the ambient ozone exposures commonly encountered throughout the forests of eastern USA (Davis and Skelly, 1992; Simini *et al.*, 1992; Skelly *et al.*, 1992; Hildebrand *et al.*, 1996). Foliar ozone-induced symptoms on broadleaved species are manifest as an upper leaf surface stipple which appears late in the growing season; early leaf reddening and leaf senescence are also commonly observed in very sensitive genotypes of black cherry. During the surveys, typical adaxial leaf surface stipple was observed on numerous indigenous capulín black cherry (41%), with injury increasing notably with increased elevation within the Park boundaries. Injury was also reported to be more severe within the capulín black cherry compared with similar survey results in central Pennsylvania (Skelly *et al.*, 1992) and within the Shenandoah National Park of Virginia (Hildebrand *et al.*, 1996). Defoliation of capulín black cherries in Desierto de los Leones reached more than 80% and, as such, the defoliation was considered to be of serious consequence to the health and productivity of this important black cherry variety in the forests proximal to the Valley of Mexico. The authors concluded that there was an obvious need for continuing observations and more detailed research on exposure–response relationships in the Desierto de los Leones forests, along with more monitoring of the various pollutants of concern to vegetation throughout the entire regional forested areas.

5.2 Atmospheric deposition

Most recently, the ozone concentrations registered in Desierto de los Leones have exceeded the air quality standard for Mexico on an almost daily basis (Quevedo *et al.*, 1999). The Valley of Mexico is also considered an important source of nitrate and sulphur, because of increasing industrial activity and the heavy traffic present in the metropolitan area. Both factors influence the whole basin (Benitez, 1992).

Based on the similarities demonstrated by Miller *et al.* (1994) between the Valley of Mexico and the San Bernardino Mountains of southern California in relation to exposure of ozone and its induced injury to pine trees, several cooperative studies have been funded. These new studies are examining the nutrient status of pine foliage and the deposition of nitrogen and sulphur compounds at the National Parks of Desierto de los Leones and Zoquiapan, considered as high and low deposition areas, respectively (Fenn *et al.*, 1999). These places, along the Ajusco mountain chain, have become special study sites for national research institutions as well as for international projects. Several recent studies have been supported by the North American Forestry Commisssion of the UN Food and Agriculture Organization.

There is some debate as to whether Mexico City should be considered as a single source of pollutants or as a regional source. In addition to ozone precursors, large quantities of nitrogen and sulphur compounds are emitted, leading to very high deposition rates for these elements. The deposition may be causing significant changes to forest ecosystems in downwind areas, e.g. the Desierto de los Leones National Park. Depending on the nature of the soil, the capacity for nitrification and the movement of nitrogen to surface waters, nitrogen in these waters within the area may increase. As reported in other areas with high nitrogen deposition, vegetation growth may be stimulated when nitrogen is a limiting factor, and when the deposition is not accompanied by elements interfering with nutrient uptake and/or where the total deposition does not introduce additional stress (Hall *et al.*, 1996).

Some investigators have found elemental deficiencies in sacred-fir foliage (López, 1996). Complex chemical reactions are probably taking place, such as impairments to nutrient uptake. Based on the work of Fenn *et al.* (1999), it is likely that atmospheric deposition of nitrogen in the downwind area has resulted in nitrogen saturation of the forest, causing significant exports of this element from watersheds. In these studies, it has also been shown that sulphur deposition (although higher at the downwind area, i.e. Desierto de los Leones National Park) is not as low as expected from measurements at a distant site (Zoquiapan) that is not directly exposed to the urban plume. In the latter case, local sulphur sources may play a significant role, such as in the vicinity of the recently active Popocatepetl volcano (see section 4.2) (Baez *et al.*, 1997).

6 Discussion and conclusions

The forests of the southern areas of the Valley of Mexico have been subjected to oxidant stress for more than 30 years as a result of their proximity to the Mexico City metropolitan area. Studies of air pollution impacts at specific sites – namely, Ajusco (the mountain corresponding to the volcano of the same name) (Bauer, 1991) and the Desierto de los Leones National Park (further to the west) – have provided valuable insights into the problem, although research on many aspects of pollutant deposition is still needed.

Since the air pollution injury on *P. hartwegii* and *P. leiophylla* was described (Krupa and Bauer, 1976), not only has the urban plume become more conspicuous, but the city itself has increased in size and many forested areas are now within the urban zone. Consequently, the surrounding forests are suffering from more severe pollutant stresses than in the past. The stress may be leading to changes in vegetation composition, as has been documented elsewhere (e.g. Miller, 1973; Miller *et al.*, 1991; Miller and McBride, 1999).

Throughout the recent years of observations, it has been repeatedly confirmed that *P. hartwegii* is the most sensitive species among the nine *Pinus* species of the mixed conifer forest type present in the mountain ranges to the south of Mexico City at altitudes from 2800 to 3500 m elevation (Martínez, 1963). Another major species, *A. religiosa*, is also highly sensitive to ozone and is located at altitudes between 2800 and 3800 m (Alvarado *et al.*, 1993); oaks, capulín black cherry, cedar and willows also occur, and some of these have been shown to be sensitive to photochemical oxidants. The responses of individual species to oxidant pollutants are variable, as demonstrated for *P. hartwegii* (Hernández, 1984) and *P. serotina* var. *capuli* (Skelly *et al.*, 1997).

P. hartwegii has almost disappeared from the highest peaks of those areas facing Mexico City. This is also the case for *A. religiosa* trees located in the windward zone exposed to the city, specifically at the distal ends of the ravines (Alvarado *et al.*, 1991). Mortality of both *P. hartwegii* and *A. religiosa* has been extensive. The death of such species has been attributed to long-term high ozone concentrations, with peak exposure concentrations up to 360 ppb. Other factors may also have played a significant negative role (Alvarado *et al.*, 1993), including but not limited to the lack of forest management, excessive water extraction leading to drought, insect pests and several possible pathogens that may contribute to and accelerate the decline process leading to the eventual death of the *A. religiosa* forest. Similar considerations are also valid for the ozone-sensitive pine species *P. hartwegii* and the broadleaved species *Prunus serotina* var. *capuli*.

Studies and observations conducted over more than 20 years indicate that vegetation changes are taking place at numerous sites exposed to continuous oxidant stress. Such changes may lead to changes in the forest composition through the disappearance of less tolerant species within the diverse strata of the forest ecosystem.

Future studies should focus on determining the sensitivity of other forest and native plant species to ambient ozone exposures, determining the exposures being encountered within the forests surrounding the mountainous areas of Mexico City, and determining long-term changes in the health of the forests. These forests are of tremendous importance to the health and well-being of the inhabitants of the Valley of Mexico. A long-term goal of this much needed research must be to work with the governing authorities and regulatory agencies to improve air quality, thereby reducing the multiplicity of problems that have recently become of such concern to many scientific communities around the world.

There is a general trend for the cities of Latin America to be actively growing, presenting the possibility that other cities may develop similar problems to those of Mexico City. A common picture is the continuous arrival of rural migrants. The tradition of centralism, which has very deep roots in the mentality of the people, together with a diverse range of government subsidies, make the bigger cities attractive, especially to the rural population. This trend is found in Mexico City and also in the Brazilian cities of Rio de Janeiro and São Paulo and in Santiago in Chile. These cities do not have a clear future, and are already fighting to survive within the context of a polluted atmosphere and severely depleted resources. It is likely that many of the lessons learned from Mexico City can be applied to other areas, and it is important that the knowledge that has been gained is applied to the solution of the environmental problems facing the cities of the region.

References

Alarcón, M.A., Bauer, L.I. de, Jasso, J., Segura, G. and Zepeda, E.M. (1995) Patrón de crecimiento radial en árboles de *Pinus hartwegii* afectados por contaminación atmosférica en el Suroeste del Valle de México. *Agrociencia, Serie: Recursos Naturales Renovables* 3(3), 67–80.

Alvarado, D. (1989) Declinación y muerte del bosque de oyamel (*Abies religiosa*) en el Sur del Valle de México. MSc thesis, Colegio de Postgraduados, Montecillo, Méx., Mexico.

Alvarado, D. and Bauer, L.I. de (1991) Ataque de *Lophodermium* sp. en poblaciones naturales de *Pinus hartwegii* de 'El Ajusco', México, bajo el efecto de gases oxidantes. *Micología Neotropical Aplicada* 4, 99–109.

Alvarado, D., Bauer, L.I. de and Galindo, J. (1991) Declinación y muerte del bosque de oyamel (*Abies religiosa*) en el Sur del Valle de México. *Agrociencia, Serie: Recursos Naturales Renovables* 1(2), 123–143.

Alvarado, D., Bauer, L.I. de and Galindo, J. (1993) Decline of sacred fir (*Abies religiosa*) in a forest park south of Mexico City. *Environmental Pollution* 80, 115–121.

Alvarez, D.E. (1996) Determinación de alteraciones a nivel citológico e histoquímico en *Abies religiosa* (H.B.K.) Schl. et Cham. del Desierto de los Leones del D. F., y su relación con el ozono. MSc thesis, Facultad de Ciencias, Universidad Nacional Autónoma de México.

Baez, A., Belmont, R., Azara, E., Torres, M. and Padilla, H. (1997) Cloudwater and precipitation chemistry at Popocatepetl volcano: a high elevation site. *Journal of Environmental Quality* 26, 253–258.

Bauer, L.I. de (1972) Uso de plantas indicadoras de aeropolutos en la Ciudad de México. *Agrociencia, Serie D* 9, 139–141.

Bauer, L.I. de (1991) La influencia urbana en el área del Ajusco. In: *Memorias del I Simposio Nacional Sobre Agricultura Sostenible.* Comisión de Estudios Ambientales, C.P.-MOA/INT, pp. 173–181.

Bauer, L.I. de and Hernández Tejeda, T. (1986) *Contaminación: Una Amenaza Para la Vegetación en México.* Talleres Gráficos del Colegio de Postgraduados, Chapingo, Méx., Mexico, 84 pp.

Bauer, L.I. de and Krupa, S.V. (1990) The Valley of Mexico: summary of observational studies on its air quality and effects on vegetation. *Environmental Pollution* 65, 109–118.

Bauer, L.I. de, Hernández Tejeda, T. and Manning, W.J. (1985) Ozone causes needle injury and tree decline in *Pinus hartwegii* at high altitudes in the mountains around Mexico City. *Journal of the Air Pollution Control Association* 35(8), 838.

Benitez, F. (1984) *Historia de la Ciudad de México*. Salvat Mexicana de Ediciones 9:117. Mexico.

Bravo, H., Perrin, G., Sosa, E. and Torres, R. (1987) *Contaminación Atmosférica por Fuentes Móviles en la Zona Metropolina de la Ciudad de México*. Tercer Encuentro Iberoamericano sobre la Ciudad de México, México DF.

Bravo, H.A. and Torres, R.J. (1995) *Trends of Air Pollution in the Metropolitan Area of Mexico City*. Sección de Contaminación Ambiental, Universidad Nacional Autonoma de México, México DF. (Mimeo.)

Bryson, R.A. and Haze, F.K. (eds) (1974) *Climates of North America*. Elsevier Science Publishing Company, New York.

Burkart, R., Marchetti, B. and Morello, J. (1995) Grandes ecosistemas de México y de Centroamérica. In: Gallopín, G.C. (ed.) *El Futuro Ecológico de un Continente*. UNC-FCE, México DF, pp. 101–163.

Calderón, E. and Hernández, B. (1987) Crecimiento actual de la población de México. *Ciencia y Desarrollo* 76, 49–66.

Castro, J.M., González, V. and Hernández Tejeda, T. (1995) Metales pesados en los suelos del Desierto de los Leones, Distrito Federal. *Ciencia Forestal* 20(77), 101–112.

Davis, D.D. and Skelly, J.M. (1992) Foliar sensitivity of eight eastern hardwood species to ozone. *Journal of Air and Soil Pollution* 62, 269–277.

Departamento del Distrito Federal and Limusa S.A. de C.V. (eds) (1997) *Testimonios. Medio Ambiente de la Ciudad de Mexico, 1994–1997*. Limusa-Noriega Editores, México DF.

Ezcurra, E. and Mazari Hiriart, M. (1996) Are megacities viable? A cautionary tale from Mexico City. *Environment* 38(1), 6–35.

Fenn, M.E., Bauer, L.I. de, Quevedo, A. and Rodriguez, C. (1999) Nitrogen and sulfur deposition and forest nutrient status in the Valley of Mexico. *Water, Air, and Soil Pollution* 113, 155–174.

Galindo, I., González, A. and Ayala, R. (1995) Emisiones de bióxido de azúfre del volcán Popocatepetl, México durante la erupción de diciembre 1994–enero 1995. In: Centro Nacional de Prevención de Desastres-UNAM (eds) *Volcán Popocatepetl: Estudios Realizados Durante la Crisis de 1994 1995*. UNAM, México DF, pp. 245–256.

Hall, J.P., Magasi, L., Carlson, L., Stolte, K.W., Niebla, E., Bauer, L.I. de, González-Vicente, C.E. and Hernández Tejeda, T. (1996) *Health of North American Forests/L'état de santé des forêts nord-américaines/Sanidad de los Bosques de América del Norte*. Natural Resources Canada, Ottawa, Ontario, Canada.

Hernández Tejeda, T. (1981) Reconocimiento y evaluación del daño por gases oxidantes en pinos y avena del Ajusco, DF. Professional thesis, Universidad Autónoma Chapingo, Méx., Mexico, 90 pp.

Hernández Tejeda, T. (1984) Efecto de los gases oxidantes sobre algunas especies del género *Pinus* nativas del Valle de México. MSc thesis, Colegio de Postgraduados. Chapingo, Méx., Mexico, 109 pp.

Hernández Tejeda, T. and Bauer, L.I. de (1984) Evolución del daño por gases oxidantes en *Pinus hartwegii* y *P. montezumae* var. *Lindleyi* en el Ajusco, DF. *Agrociencia* 56, 183–194.

Hernández Tejeda, T. and Bauer, L.I. de (1986) Photochemical oxidant damage on *Pinus hartwegii* at the 'Desierto de los Leones', DF. *Phytopathology* 76(3), 377.

Hernández Tejeda, T. and Nieto, C. (1996) Effects of oxidant air pollution on *Pinus maximartinezii* Rzedowsky in the Mexico City region. *Environmental Pollution* 92, 79–83.

Hernández Tejeda, T., Krupa, S.V., Pratt, G.C. and Bauer, L.I. de (1981) Sensibilidad de plantulas de eucalipto (*Eucalyptus globulus* Labill.) al ozono. *Agrociencia* 43, 89–95.

Hernández Tejeda, T., Bauer, L.I. de and Ortega, M.L. (1986) Identificación y determinación de los principales pigmentos fotosintéticos de hojas de *Pinus hartwegii* afectadas por gases oxidantes. *Agrociencia* 66, 71–82.

Hildebrand, E.S., Skelly, J.M. and Fredericksen, T.S. (1996) Foliar response of ozone sensitive hardwood tree species from 1991 to 1993 in the Shenandoah National Park, Virginia. *Canadian Journal of Forest Research* 26, 658–669.

Husar, R.B. (1985) Chemical climate of North America: sources and deposition with special emphasis of high elevation locations. In: *Proceedings of the Conference on Air Pollutants Effects on Forest Ecosystems.* University of Minnesota, St Paul, Minnesota, pp. 5–38.

INE (Instituto Nacional de Ecología) (1995) *Elementos de Política Ambiental Para Una Ciudad Sustentable: Manejo de la Cuenca Atmosférica de México.* SEMARNAP México DF, 13 pp.

Jáuregui, E. (1958) El aumento de la turbiedad del aire en la Ciudad de México. *Revista Ingeniería Hidráulica en México* 12(3), w/p.

Jáuregui, E. (1983) Visibility trends in México City. *Erdkunde* 37, 296, 299.

Krupa, S.V. and Bauer, L.I. de (1976) La ciudad daña los pinos del Ajusco. *Panagfa* 4(31), 5–7.

López, M.A. (1993) Evaluación nutrimental de *Abies religiosa* en el Desierto de los Leones DF. MSc thesis, Colegio de Postgraduados Montecillo, México, 101 pp.

Martínez, M. (1963) *Las Pinaceas Mexicanas*, 3rd edn. Instituto de Biología, UNAM, México.

Martínez Gonzalez, L. and Chacalo Hilu, A. (1994) *Los Arboles de la Ciudad de México.* UAM/Unidad Azcapotzalco, México DF.

Middleton, J.T. and Haagen-Smit, A.J. (1961) The occurrence, distribution and significance of photochemical air pollution in the United States, Canada and Mexico. *Journal of the Air Pollution Control Association* 11(3), 129–134.

Miller, P.R. (1973) Oxidant-induced community change in a mixed conifer forest. In: Naegele, J.A. (ed.) *Air Pollution Damage to Vegetation.* Advances in Chemistry Series 122, 101–117.

Miller, P.R. and McBride, J.R. (eds) (1999) *Oxidant Air Pollution Impacts in the Montane Forests of Southern California.* Ecological Studies 134, Springer Verlag, New York.

Miller, P.R., Schilling, S.L. and McBride, J.R. (1991) Chronic ozone injury and associated stresses affect relative competitive capacity of species comprising the California mixed conifer forest type. In: *Memorias del I Simposio Nacional Sobre Agricultura Sostenible.* Comisión de Estudios Ambientales, C.P.-MOA/INT, pp. 161–172.

Miller, P.R., Bauer, L.I. de, Quevedo, A. and Hernández Tejeda, T. (1994) Comparison of ozone exposure characteristic in forested regions near Mexico City and Los Angeles. *Atmospheric Environment* 28, 141–148.

Miller, P.R., Stolte, K.W., Duriscoe, D.M. and Pronos, J. (technical coordinators) (1996) *Evaluating Ozone Air Pollution Effects on Pines in the Western United States.* General

Technical Report PSW-GTR-155. Pacific Southwest Research Station, Forest Service, US Department of Agriculture, Albany, New York.
Millon, R. (1970) Teotihuacan: completion of a map in the giant ancient city of the Valley of Mexico. *Science* 170, 1077–1082.
Oceano (1993) *Atlas y Geografía Universal*. Ediciones Oceano, Barcelona, Spain.
Pick, J.B. and Butler, E.W. (1997) *Mexico Megacity*. Westview Press, Boulder, Colorado.
Quadri, G. and Sánchez, L.R. (1994) *La Ciudad de México y la Contaminación Atmosférica*. Limusa-Noriega, México DF.
Quevedo, A., Bauer, L.I. de, Miller, P.R. and Perea, V. (1999) La probable relación entre algunos factores meteorológicos y la declinación forestal en el Parque Nacional Desierto de los Leones. *Ciencia Forestal en México* (in press).
Romero, H.M. (1987) *Historia del Transporte en la Ciudad de México*. México DF.
Rosas, I., Belmont, R. and Jáuregui, E. (1995) Seasonal variation of atmospheric lead levels in three sites in Mexico City. *Atmósfera* 8, 157–168.
Rzedowski, J. (1978) *Vegetación de México*. Limusa, México DF.
SARH (Secretaria de Agricultura y Recursos Hidraúlicos) (1994a) *The Forests of Mexico*. AMAC Impresos, México DF.
SARH (Secretaria de Agricultura y Recursos Hidraúlicos) (1994b) *Inventario Nacional Periódico 1992–1994*. ATT Diseño, Mexico DF.
Simini, M.J., Skelly, J.M. and Davis, D.D. (1992) Hardwood seedling response to ambient ozone exposures in northcentral Pennsylvania via open-top chambers. *Canadian Journal of Forest Research* 22, 1789–1799.
Skelly, J.M., Snyder, K.R. and Savage, J.E. (1992) Ozone induced effects to half-sib families of black cherry. *Phytopathology* 82, 247.
Skelly, J.M., Savage, J.E., Bauer, L.I de and Alvarado, D. (1997) Observations of ozone-induced foliar injury on black cherry (*Prunus serotina*, var. *capuli*) within the Desierto de los Leones National Park, Mexico City. *Environmental Pollution* 95, 155–158.
Stakman, E.C. and Christensen, J.J. (1946) Aerobiology in relation to plant disease. *Botanical Reviews* 12, 205–253.
Vazquez, J. (1986) *El Saneamiento y la Limpia Forestales en el Desierto de los Leones*. COCODER-DDF, México DF.
Velasco Levy, A. (1983) La contaminación atmosférica en la Ciudad de México. *Ciencia y Desarrollo* 9(52), 59–68.
Watmough, S.A. (1997) Heavy metals in soils and bark of sacred fir in Desierto de los Leones Park. Internal Report. Trent University, Peterborough, Ontario, Canada.
Wirth, C.J. (1997) Transportation policy in Mexico City. The politics and impacts of privatization. *Urban Affairs Review* 33(2), 155–181.

Regional Impacts, Consequences and Policy Options in Relation to Air Pollution in Latin America

4

R. Herrera* **and N. Chacón**

Centro de Ecología, Instituto Venezolano de Investigaciones Científicas, Caracas, Venezuela

Regional impacts of air pollution in Latin America are described. Although the subject has been taken into account only recently, a substantial body of research is being carried out in several countries of the region, especially in areas of rapid industrial and urban development. There are very few examples of systematic studies of inventory sources or deposition rates and even fewer on the possible impacts of air pollution on ecosystems. In large cities and their surroundings, some efforts have been implemented in relation to health problems and damage to exposed materials. Impacts on crops have been studied in a very few places downwind of large industrial or urban centres; much less is known about the impact on forests and natural ecosystems. Despite the importance of the forests of the region, the impact of air pollution on forests is largely unknown and very little work has been done on either the acute or chronic impacts of air pollution. Significant problems could be expected in some areas with aquatic ecosystems where biodiversity could be reduced.

An effort to predict the future impacts using models based on energy and development scenarios is reviewed. Some of these models predict that in areas in Brazil, Colombia, Venezuela, Chile and Ecuador, which already have localized problems, there could be significant expansion of affected areas by 2025, with deposition values reaching beyond what would be considered as high deposition rates. Coordinated efforts to increase research on air pollution are badly needed. In particular, comprehensive

* Present address: Departamento de Química, Universidad Simón Bolívar, Caracas, Venezuela.

inventories should be developed. As the process of industrialization is more recent than in Europe and North America and in many cases lags behind parts of Asia, there are options for choosing development patterns that would minimize the risks of air pollution impact. Recent efforts to reach and influence policy makers in Latin America are also reviewed.

1 Introduction

In Latin America, air pollution problems and some of their consequences, such as acid rain, have been associated with urban environments and industrial centres. Despite the growing concern of both governments and citizens, few systematic studies of the sources of air pollution have been undertaken. Air quality monitoring networks, together with research on impacts of air pollution on health, crops, ecosystems and materials, are notoriously lacking. In 1986, a Workshop on Acidification in Tropical Countries was organized by the Scientific Committee on Problems of the Environment (SCOPE) in Venezuela. At this event, experts from industrialized countries as well as from developing countries in Asia, Africa and Latin America met to study the state of the art on air pollution and its likely effects in tropical countries (Rodhe and Herrera, 1988). In 1996, the Stockholm Environment Institute (SEI) organized three workshops for regional experts to evaluate systematically the sensitivity of terrestrial ecosystems to acidic deposition in Asia, Africa and Latin America (Kuylenstierna *et al.*, 1998). Rodhe *et al.* (1988) considered that in the lowlands of South America, in areas of high rainfall with naturally acidic soil of low exchange capacity, ecosystems could be highly sensitive to increased acidification. In the case studies in Brazil (Moreira-Nordemann *et al.*, 1988) and Venezuela (Sanhueza *et al.*, 1988), they considered that, starting from urban areas and industrial centres such as in north-east South America and south-east Brazil, further acidification could occur.

In order to evaluate the risks created by increased acidification, it is necessary to inventory sources systematically, to assess ecosystem sensitivity, and finally to evaluate social costs *vis-à-vis* possible development scenarios. In the workshop held in Costa Rica in October 1996 experts from Argentina, Brazil, Costa Rica, Chile, Mexico, Peru and Venezuela met with other specialists to evaluate the sensitivity of ecosystems using sensitivity maps for acidic deposition prepared using geographic information systems (GIS). Sanhueza (1992), in a study of changes in tropospheric chemistry, indicated that although short-lived gases in the lower atmosphere (precursors of acidification) were mostly concentrated in the northern hemisphere, this problem could also develop in the southern hemisphere associated with industrial development. More recently, Sanhueza (1997) and Sanhueza and Santana (1994) assessed wet and dry deposition in South America and Central America. They found that, away from large cities and industrial centres, there are are relatively low deposition rates for nitrate (NO_3), ammonium (NH_4) and sulphate (SO_4); while in densely populated areas such as Caracas and Valencia in Venezuela, and Rio de Janeiro, São Paulo and São José dos Campos in Brazil, they found concentrations similar to those in rural areas in Europe and North America, where there are already visible effects of acidification.

In addition to the known effects of industrial activities on acidification processes, the growing number of cars and the general increase in energy use associated with development and population growth, in Latin America there are also air pollutants originating from biomass burning (Fig. 4.1). The utilization of fuelwood, the use of fire (e.g. for the burning of sugarcane residues, as a means for deforestation, or as a practice to renew pastures) and the large-scale uncontrolled vegetation fires, such as in Roraima in the Brazilian Amazon in 1998, produce large amounts of air pollution. Crutzen and Andreae (1990) estimated that biomass burning in the tropics can reach equivalent emissions of 1800 to 4700 Tg C per year (Tg = 10^{12} g), with other emissions ranging from 2.1–5.5 Tg nitrogen oxides (NO_x), 0.5–2.0 Tg ammonia (NH_3), 0.1–0.3 Tg nitrous oxide (N_2O) and from 1.0–4.0 Tg SO_4. Amongst both the general population and the scientific community, air pollution is perceived in Latin America as a problem associated with large cities and industrial areas and it has not yet received prominence on political agendas. Warning systems have been implemented in some large cities in Latin America, with cities such as Mexico City and Santiago de Chile (Fig. 4.2) imposing restrictions on the use

Fig. 4.1. Smoke generated by burning forests near Merida, Venezuela.

Fig. 4.2. Poor air quality in Santiago de Chile.

of motor vehicles, and there have been temporary closures of industries and policies to move industries away from cities.

In Latin America, it seems that we are a step behind some regions of Asia where serious local air pollution problems have already become regional (see Chapters 7, 8 and 9). In this context we must consider two important factors. Firstly, there is a possibility that air pollution could soon reach rural areas and even have transboundary implications in some cases. Secondly, as a result of the natural buffering capacity of ecosystems, the effects of accumulated acidification could go unnoticed until the critical load is exceeded, and the effects of acidification would only then be noticed.

2 Direct impacts of air pollutants in the countries of Latin America

2.1 Impacts on human health and exposed materials

In studies in Santiago de Chile and Mexico City, a close correlation between the deterioration of human health and ambient ozone (O_3) concentrations has been found. In Chile, Ostro *et al.* (1996) found that a daily increase of 10 μg m^{-3} in ambient ozone was associated with a 1% increase in mortality. Nasal inflammation and other responses in children have been associated with ozone exposure in Mexico City (Calderongarcidueñas *et al.*, 1995). Field studies have shown that eye, nose and throat irritation, chest congestion, coughing and headache can be associated with average exposures to ozone concentrations of 200 $\mu g\ m^{-3}$ or more.

The direct effects of gaseous pollutants on materials, buildings and the cultural heritage should be studied in Latin American countries as many of them are in warm and humid regions with a high frequency of precipitation, all factors that increase the potential damage of acidic pollutants to materials that, in turn, implies higher maintenance costs, replacement or irreparable losses to historical monuments. In Latin America, the increase in SO_2 concentrations and increasing numbers of automobiles are becoming threats to man-made materials. In these countries it is important to plan emission reductions using inventories, pollution levels and corrosion maps for each region.

2.2 Impacts on crops and forests

Few exposure–response studies have been carried out in Latin America with crops and forests. Rao *et al.* (1993) studied the responses of eight tropical plants and their relative tolerance to SO_2 deposition through increases in nitrogen uptake. A few other isolated case studies have shown effects on crop production. In the Valley of Mexico up to 40% losses in beans have been attributed to ozone exposure (Laguette-Rey, 1986). There is also clear evidence of forest decline in the mountains around Mexico City, where ozone exposure reaches values comparable to those around Los Angeles (Tovar, 1989; see also Chapter 3). The impacts of nitrogen deposition are less understood, mainly because little is known about the critical loads for this pollutant in Latin American ecosystems; more research is needed on risk assessment for nitrogen deposition in tropical and sub-tropical ecosystems.

3 Risk assessment in Latin America

Even though the experimental evidence mentioned above demonstrates the effects of air pollution on various crops and forests in several places in Latin America, and especially highlights the effects of the increases in ozone concentrations, the available studies remain few and far between. Consequently, risk assessment studies are needed to achieve a better geographical coverage of investigations into the nature and degree of air pollution impacts in Latin America. Almost all areas in developing nations with a moderate to high risk of damage by ozone are in agricultural zones. In Latin America, field studies have shown detrimental effects of ozone in croplands in Mexico and Venezuela (Ashmore and Marshall, 1998).

3.1 Risks of air pollution in Brazil

Although global assessments are beginning to identify the general magnitude of the problem and identify priority areas for research, national studies in each region are badly needed. In Latin America, more detailed studies have been carried out in Brazil, a large country with a variety of combinations of pollution sources, population density and agricultural systems. Crops like soybean, tomato, potato and cotton are usually grown in the vicinity of large population centres, such as the south-east of São Paulo State, Rio de Janeiro and Minas Gerais (Ashmore and Marshall, 1998). The seasonal burning of sugarcane fields, used to produce both sugar and ethanol for car fuel, has been shown to produce increases in ozone concentrations; the same effect has been observed during the seasonal burning of forests (Kirchoff *et al.*, 1991, 1992).

It is difficult to assess the risks of air pollution damage in the rest of the country because of the paucity of data on gaseous pollutants. Agricultural data are more consistent and widespread and could be used to identify areas at risk from damage to crops and other ecosystems based on global emission estimates.

4 Ecosystem acidification

Chemical composition studies of total precipitation (including wet and dry deposition) in Latin America indicate that in remote areas such as the Amazon rainforests, biogenic organic acids are responsible for the low pH of precipitation, while strong acids such as nitric and sulphuric acid represent 10–20% of total hydrogen ions (Andreae *et al.*, 1988, 1990; Lesak and Melack, 1991; Williams *et al.*, 1997). In savannah areas in Venezuela, acidic precipitation has also been found to be related to biogenic organic acids emitted during seasonal burning of the vegetation (Montes *et al.*, 1985; Sanhueza *et al.*, 1987, 1989, 1991; Sanhueza, 1992). Near population or industrial centres, this pattern of acidic precipitation is altered and inorganic acids such as sulphuric and nitric acid and, to a lesser extent, hydrochloric acid are responsible for over 40% of total acidity (Lewis, 1981; Lewis and Weibezahn, 1981; Jickells *et al.*, 1982; Hendry *et al.*, 1984; Da Silva Filho *et al.*, 1987; De Mello *et al.*, 1987; Galloway *et al.*, 1989; Hermoso *et al.*, 1990; Gordon *et al.*, 1994; Morales *et al.*, 1995; Baez *et al.*, 1997).

In remote areas, away from pollution sources, precipitation acidity is mainly caused by dissolved CO_2 and organic acids of biogenic origin, and pH is typically around 5 (Sanhueza and Santana, 1994); near urban centres precipitation is lower, with the pH varying between 4 and 5 (Gordon *et al.*, 1994). Although these levels of acidity are lower than those found in near-pristine areas in Europe and North America, a strong tendency for decreased pH levels is currently being reported as being caused by both organic acid emissions through biomass burning and increases in emissions of acidic precursors in urban and industrial centres.

The impacts of acidification in Latin America are still little known but it is worth noting that, in general, emissions of nitrogen and sulphur compounds increase as the population increases and industry expands. Consequently, SO_2 and NO_x emissions could exceed present global estimates (Galloway, 1989) and thus increase the threat of acidification to different ecosystems of the region.

4.1 Impacts of acidity on forest vegetation, crops, biodiversity and water bodies

The best example of pollutant-induced forest damage in South America is in the Serra do Mar rainforest, around Cubatão, in São Paulo State, Brazil. Nutrient cycles have been interrupted by high leaching losses of nitrogen and calcium and the forest has been severely disturbed at several sites by extremely high pollution loads (Mayer *et al.*, 1996). The pollutants involved are primarily SO_2 and hydrogen fluoride, but heavy metals and other pollutants may also be impacting the vegetation.

The majority of soils in the tropics are naturally acidic, but cultivated soils are not, either because of better soil quality selection or as a result of liming. Population increases and the scarcity of agricultural soils mean that acidic soils are frequently cultivated (Sanchez, 1976), resulting in low yields. The poor crop development on these soils is caused by high aluminium concentrations and/or low base saturation. There are distinct responses to acidity for different crops, depending on their tolerance to high aluminium or manganese levels or their requirements for calcium and magnesium. Crops such as pineapple, coffee, tea, rubber and melon are tolerant of high values of exchangeable aluminium but are not necessarily tolerant of other ions such as manganese (Sanchez, 1976).

The impact of acidity on biodiversity in developing countries is still poorly understood. The diversity of some tropical habitats is so high that changes are extremely difficult to detect and interpret (McDowell, 1988), especially as many of the organisms remain undescribed. For some animal species, however, the impact could be significant. As an example we could cite amphibians in tropical rainforests which, being an integral part of these ecosystems, run a high risk because they are exposed to small changes in rain pH (McDowell, 1988). There have recently been unexplained disappearances of some amphibian species in tropical forests (e.g. Weygoldt, 1989; Pounds *et al.*, 1997; Lips, 1999) but the role of acidification remains very speculative, and it seems likely that the fungal disease chytridiomycosis is partly or wholly responsible for the declines of some of the species in Australia, Central America and South America (Berger *et al.*, 1998; Pessier *et al.*, 1999).

6 Deposition and transfer in the atmosphere

The transfer and deposition of air pollutants (SO_2, NO_x, NH_3 and O_3) have been studied through mathematical models. The results of the MOGUNTIA models for 1985 (Rodhe *et al.*, 1995) show that in the northern hemisphere sulphur deposition is high only in areas near the sources, while for 2050 a substantial increase in deposition is likely in certain areas in Latin America, especially Mexico and Brazil. Regional modelling of sulphur deposition using the ACTM (Atmospheric Chemistry Transfer Model) indicates that in 1990 sulphur deposition was low for South America, with some areas of high deposition around the larger cities of Brazil, Venezuela, Colombia, Peru, Chile and Argentina. For 2025 the model predicts an increase in the size of areas of high deposition and for 2050 areas of high deposition (over 1000 mg S m^{-2}) would cover significant areas in south-east Brazil, the surroundings of Buenos Aires, and northern Colombia and Venezuela. By 2050, deposition around cities in Chile, Peru and Ecuador may reach values as high as in 2025 but is unlikely to affect extensive areas in these countries.

For nitrogen, the models predict an increase in deposition by 2020 for the South American region, using fuel-burning rates linked to population increase and per capita consumption of energy. According to the STOCHEM model, ozone concentrations in developing countries are likely to increase between 1992 and 2015. High concentrations would occur for South America during August, especially in Brazil.

7 Risk assessment at global and regional scales

Using global estimates for sulphur deposition for 1985, there was no appreciable risk of regional-scale problems in Latin America for that year, except for isolated areas. By 2050, the scenario is very different, with a much higher risk of acidification impacts, especially in Brazil and Mexico. Areas of significant risk can also be identified in Colombia, Ecuador, Peru and Venezuela.

Regional-scale risk assessments for Latin America for the scenarios of 1990, 2025 and 2050 suggest that the differences between global and regional scales have greater differences than the models for Asia, especially for 2050. These differences can be explained by factors related to the use of scales. At the regional scale in Latin America for 1990, there are areas of limited risk around Rio de Janeiro, in southern Brazil and in Colombia. For the years 2025 and 2050 the area of risk increases in Brazil and Colombia and new areas appear around the larger cities in Venezuela, Peru, Ecuador, Chile and some places in the Caribbean.

8 Options to reduce and control emissions

In Latin America, as in the rest of the world, there is a growing awareness that energy-intensive development schemes are inevitably linked to an increase in acidifying emissions. Experience in Europe has shown that even though high emission rates were reached in 1980, environmental policy decisions resulted in emissions being drastically reduced without compromising development and economic growth. There is also evidence that high rates of economic growth can be achieved even with decreases in energy use intensity (Jackson, 1997).

In Latin America, where there is ample room for planning, it is still possible to formulate development policies that move away from the obsolete industrialization models that led many developed countries to pay a high environmental price as a result of air pollution. Among the available options, there are control measures that if applied to industries and power plants would reduce emissions, especially if cleaner fuels were used, more modern combustion methods were adopted or systems were installed at the flue end of combustions to reduce pollutants. For Latin America, the emission levels could be reduced by almost half by 2025 and would be below 1990 levels by 2050. Preventive approaches result in more positive scenarios of future emissions, and have considerable potential in Latin America. A first step would be an increase in the efficiency of energy use, both in the industrial sector and in transport. In electricity generation, there is huge potential for improvements in efficiency, typically of the order 30–35%, and up to 50% in the case of combined cycle turbines. The substitution of conventional energy sources such as gasoline or diesel for sources such as natural gas or alcohol is already practised in several countries, and is resulting in important reductions of pollutant emissions. Some countries in Latin America have very considerable hydroelectric potential, and utilization of this together with the development of other renewable sources could result in lower pollution levels. However, any such hydroelectric developments would need to take into account other environmental values, creating the need to find an optimal solution. The International Shell Petroleum Company estimates that, at the global level, renewable sources of energy could provide up to 50% of primary energy by 2050.

In 1998, a policy dialogue was held in Buenos Aires, Argentina, hosted by the International Institute for Environment and Development–Latin America and organized by the Fundación Futuro Latinoamericano (FFLA); the meeting was attended by delegates from the four MERCOSUR countries (Argentina, Brazil, Paraguay and Uruguay) and a wide range of representatives from sectors including the trade unions, media, non-governmental organizations and policy makers. The Cañuelas Declaration (see Chapter 12) was signed on the Control and Prevention of Atmospheric Pollution. This Declaration was followed by agreements at ministerial levels at a meeting of Ministers of the Environment of MERCOSUR.

9 Conclusions

Latin America is characterized by rapid development and population expansion. This has already resulted in problems associated with decreasing air quality (see Chapter 3 for the example of Mexico City) and problems also exist in other cities, such as Santiago (Chile) and São Paulo (Brazil). Impacts on forests remain largely unknown, and very little work has been done on either the acute or the chronic impacts of air pollution in vegetation in Latin America. Given the importance of the forests of the region in the maintenance of biodiversity and other environmental services, this gap in our knowledge is important.

The prospects in Latin America to achieve and enforce agreements of the type described above seem promising as transboundary pollution problems are just emerging, development models allow a wider choice and population density problems are concentrated in urban areas rather than being more widely distributed.

Acknowledgements

A version of this paper was published in Spanish by the Stockholm Environment Institute (SEI) as a background document for the MERCOSUR Policy Dialogues held in Argentina in June 1998 (Kuylenstierna *et al.*, 1998).

References

Andreae, M.O., Talbot, R.W., Andreae, T.W. and Harriss, R.C. (1988) Formic and acetic acids over the Central Amazon region, Brazil. I. Dry season. *Journal of Geophysical Research* 93, 1616–1624.

Andreae, M.O., Talbot, R.W., Berresheim, H. and Beecher, K.M. (1990) Precipitation chemistry in Central Amazonia. *Journal of Geophysical Research* 95, 16987–16999.

Ashmore, M.R. and Marshall, F.M. (1998) Direct impacts of pollutant gases on crops and forest. In: Kuylenstierna, J. and Hicks, K. (eds) *Regional Air Pollution in Developing Countries, Background Document for Policy Dialogue, Bangkok, March 1998.* Stockholm Environment Institute, York, pp. 21–33.

Baez, A.P., Azpra, B.E., Torres, M.B. and Padilla, H.G. (1997) Cloudwater and precipitation chemistry at Popocatepetl Volcano: a high elevation site. *Journal of Environmental Quality* 26, 253–258.

Berger, L., Speare, R., Daszak, P., Green, D.E., Cunningham, A.A., Goggin, C.L., Slocombe, R., Ragan, M.A., Hyatt, A.D., McDonald, K.R., Hines, H.B., Lips, K.R., Marantelli, G. and Parkes, H. (1998) Chytridiomycosis causes amphibian mortality associated with population declines in the rain forests of Australia and Central America. *Proceedings of the National Academy of Sciences, USA* 95, 9031–9036.

Calderongarcidueñas, L., Rodiquezalcaras, A., Garcia, R., Ramirez, L. and Barragan, G. (1995) Nasal inflammatory responses in children exposed to a polluted urban atmosphere. *Journal of Toxicology and Environmental Health* 45, 427–437.

Crutzen, P.J. and Andreae, M.O. (1990) Biomass burning in the tropics: impact on atmospheric chemistry and biogeochemical cycles. *Science* 250, 1669–1678.

Da Silva Filho, E.V., Coelho Ovalle, A.R. and Brown, I.F. (1987) Precipitação ácida no Parque Nacional da Tijuca, Río de Janeiro. *Ciencia e Cultura* 39, 419–422.

De Mello, W.Z., De Souza, M. and Thompson Motta, J. (1987) Medidas dos fluxos de deposição seca em Niteroi, RJ – nota preliminar. *Ciencia e Cultura* 39, 1075–1078.

Do Espírito Santo, C.M. and Moreira-Nordemann, L.M. (1988) Un modelo simplificado para estimativa da taxa de percipitação do SO_2 antropogenico an Grande São Paulo. *Ciencia e Cultura* 40, 78–82.

Galloway, J.N. (1988) Effects of acid deposition on tropical aquatic ecosystems. In: Rodhe, H. and Herrera, R. (eds) *Acidification in Tropical Countries*. SCOPE Report 36, John Wiley & Sons, Chichester, pp. 141–164.

Galloway, J.N. (1989) Atmospheric acidification: projections for the future. *Ambio* 18, 161–166.

Galloway, J.N., Keene, W.C., Artz, R.S., Miller, J.M., Church, T.M. and Knap, A.H. (1989) Processes controlling the concentration of SO_4, NO_3, NH_4, H, HCOO, and CH_3COO in precipitation on Bermuda. *Tellus* 41(B), 427–443.

Gordon, C.A., Herrera, R. and Hutchinson, T.C. (1994) Studies of fog events at two cloud forests near Caracas, Venezuela. II. Chemistry of fog. *Atmospheric Environment* 28, 323–337.

Hendry, C.D., Berish, C.W. and Edgerton, E.S. (1984) Precipitation chemistry at Turrialba, Costa Rica. *Water Resources Research* 20, 1677–1684.

Hermoso, M., Santana, M. and Sanhueza, E. (1990) Características ácido-básicas de las lluvias de un sitio urbano y otro sub-urbano de Caracas. *Acta Científica Venezolana* 41, 191–198.

Jackson, T. (1997) *Power in Balance. Energy Challenges for the 21st Century*. Friends of the Earth, London.

Jickells, T., Knap, A., Church, T., Galloway, J. and Miller, J. (1982) Acid rain on Bermuda. *Nature* 297, 55–57.

Johnson, N. and Parnell, R.A. (1986) Composition, distribution and neutralization of 'acid rain' derived from Masaya volcano Nicaragua. *Tellus* 38(B), 106–117.

Kato, N. and Akimoto, H. (1992) Anthropogenic emissions of SO_2 and NO_x in Asia: emission inventories. *Atmospheric Environment* 26A, 2997–3017.

Keller, M., Kaplan, W.A. and Wofsy, S.C. (1986) Emissions of N_2O, CH_4 and CO_2 from tropical forest soils. *Journal of Geophysical Research* 91, 11791–11802.

Kirchoff, V.W.J.H., Marinho, E.V.A., Dias, P.L.S., Calheiros, R., Andre, R. and Volpe, C. (1991) Enhancements of CO and ozone from burning in sugar cane fields. *Journal of Atmospheric Chemistry* 12, 87–102.

Kirchoff, V.W.J.H., Nakamura, Y., Marinho, E.V.A. and Mariano, M.M. (1992) Excess ozone production in Amazonia from large-scale burnings. *Journal of Atmospheric and Terrestrial Physics* 54, 583–588.

Kuylenstierna, J., Hicks, K. and Herrera, R. (eds) (1998) *Contaminación Atmosférica Regional en los Países en Desarrollo*. Stockholm Environment Institute, York.

Kuylenstierna, H.W., Vallack, S. and Cinderby, S. (1998) Potential emission increases in developing countries. In: Kuylenstierna, J. and Hicks, K. (eds) *Regional Air*

Pollution in Developing Countries, Background Document for Policy Dialogue, Bangkok, March 1998. Stockholm Environment Institute, York, pp. 61–71.

Laguette-Rey, H.D., Bauer, L.I., Kohashi Shibata, J. and Mendoza, N.M. (1986) *Impacto de los Oxidantes Ambientales en el Cultivo de Frijol, en Montecillos, Estado de Mexico*. Centro de Fitopatologia, Mexico City.

Lesack, L.F. and Melack, J.M. (1991) The deposition, composition, and potential sources of major ionic solutes in rain of the Central Amazon basin. *Water Resources Research* 27, 2953–2977.

Lewis, W.M. (1981) Precipitation chemistry and nutrient loading by precipitation in a tropical watershed. *Water Resource Research* 17, 169–181.

Lewis, W.M. and Weibezahn, F. (1981) Acid rain and major seasonal variation of hydrogen ion loading in a tropical watershed. *Acta Científica Venezolana* 32, 236–238.

Lips, K.R. (1999) Mass mortality and population declines of anurans at an upland site in western Panama. *Conservation Biology* 13, 117–125.

Mayer, R., Liess, S., Silveira Lopes, M. and Kreutzer, K. (1996) *Air Pollution and Vegetation Damage in the Tropics – The Sierra do Mar as an Example*. Background document for the sensitivity map validation workshop for Latin America, Costa Rica, 1996. Stockholm Environment Institute, York.

McDowell, W.H. (1988) Potential effects of acid deposition on tropical terrestrial ecosystems. In: Rodhe, H. and Herrera, R. (eds) *Acidification in Tropical Countries*. SCOPE Report 36, John Wiley & Sons, Chichester, pp. 117–140.

Montes, R., San José, J.J., García-Miragaya, J. and Parra, N. (1985) pH of bulk precipitation during 3 consecutive annual courses in the trachypogon savannas of the Orinoco Llanos, Venezuela. *Tellus* 37(B), 304–307.

Morales, J., Bifano, C. and Escalona, A. (1995) Rainwater chemistry at the western savannah region of the Lake Maracaibo basin, Venezuela. *Water, Air, and Soil Pollution* 85, 2325–2330.

Moreira-Nordemann, L.M., Forti, M.C., Di Lascio, V.L., Monteiro, C. and Danelon, O.M. (1988) Acidification in south-eastern Brazil. In: Rodhe, H. and Herrera, R. (eds) *Acidification in Tropical Countries*. John Wiley & Sons, Chichester, pp. 258–296.

Ostro, B., Sanchez, J.M., Aranda, C. and Eskeland, G.S. (1996) Air pollution and mortality – results from study of Santiago, Chile. *Journal of Exposure Analysis and Environmental Epidemiology* 6, 97–114.

Pessier, A.P., Nichols, D.K., Longcore, J.E. and Fuller, M.S. (1999) Cutaneous chytridiomycosis in poison dart frogs (*Dendrobates* spp.) and White's tree frogs (*Litoria caerulea*). *Journal of Veterinary Diagnostic Investigation* 11, 194–199.

Pounds, J.A., Fogden, M.P.L., Savage, J.M. and Gorman, G.C. (1997) Tests of null models for amphibian declines on a tropical mountain. *Conservation Biology* 11, 1307–1322.

Rao, M., Khujneri, S., Dubey, P.S. and Kumawat, D.M. (1993) Response of eight tropical plants to enhanced ammonia deposition under field conditions prevalent with SO_2 and NH_3. *Water, Air, and Soil Pollution* 71, 331–345.

Rodhe, H. and Herrera, R. (eds) (1988) *Acidification in Tropical Countries*. SCOPE Report 36, John Wiley & Sons, Chichester.

Rodhe, H., Cowling, E., Galbally, I., Galloway, J.N. and Herrera, R. (1988) Acidification and regional air pollution in the tropics. In: Rodhe, H. and Herrera, R. (eds)

Acidification in Tropical Countries. SCOPE Report 36, John Wiley & Sons, Chichester, pp. 3–39.

Rodhe, H., Langner, J., Gallardo, L. and Kjellstrom, E. (1995) Global scale transport of acidifying pollutants. *Water, Air, and Soil Pollution* 85, 37–50.

Romero, J. and Sanhueza, E. (1986) Preliminary evaluation of SO_2 and NO_x emissions from natural and anthropogenic sources in Venezuela. In: *XXXVI Convención Anual de AsoVAC, Noviembre 1986, Universidad de Carabobo, Valencia-Venezuela.* Universidad de Carabobo, Valencia, Venezuela.

Sanchez, P.A. (1976) *Properties and Management of Soils in the Tropics.* John Wiley & Sons, New York.

Sanhueza, E. (1992) Global and hemispheric changes in tropospheric chemistry. *Interciencia* 17, 208–217.

Sanhueza, E. (1997) Deposition in South and Central America. In: Whelpdale, D.M. and Kaiser, M.S. (eds) *Global Acid Deposition Assessment.* Global Atmospheric Watch Publication No. 106, World Meteorological Office, Geneva, pp. 135–144.

Sanhueza, E. and Santana, M. (1994) Atmospheric wet deposition in Tropical America. *Israel Journal of Chemistry* 34, 327–334.

Sanhueza, E., Graterol, N. and Rondón, A. (1987) Rainfall pH in the Venezuelan savannah. *Tellus* 39(B), 329–332.

Sanhueza, E., Cuenca, G., Gómez, M.J., Herrera, R., Ishizaki, C., Martí, I. and Paolini, J. (1988) Characterization of the Venezuelan environment and its potential for acidification. In: Rodhe, H. and Herrera, R. (eds) *Acidification in Tropical Countries.* SCOPE Report 36, John Wiley & Sons, Chichester, pp. 197–255.

Sanhueza, E., Elbert, W., Rondón, A., Arias, M. and Hermoso, M. (1989) Organic and inorganic acids in rain from a remote site of the Venezuelan savannah. *Tellus* 41(B), 170–176.

Sanhueza, E., Ferrer, Z., Romero, J. and Santana, M. (1991) HCHO and HCOOH in tropical rains. *Ambio* 20, 115–118.

Spiro, P.A., Jacob, D.J. and Logan, J.A. (1992) Global inventory of sulphur emissions with a 1 × 1 degree resolution. *Journal of Geophysical Research* 97, 6023–6026.

Tovar, M. (1989) Air pollution and forest decline near Mexico City. *Environmental Monitoring and Assessment* 12, 49–58.

Várhelyi, G. (1985) Continental and global sulphur budgets. I. Anthropogenic SO_2 emissions. *Atmospheric Environment* 19, 1029–1040.

Veldkamp, E. and Keller, M. (1997) Nitrogen oxide emissions from a banana plantation in the humid tropics. *Journal of Geophysical Research* 102, 15889–15898.

Weygoldt, P. (1989) Changes in the composition of mountain stream frog communities in the Atlantic Mountains of Brazil: frogs as indicators of environmental deteriorations? *Studies of Neotropical Fauna and Environments* 24, 249–255.

Williams, M.R., Fisher, T.R. and Melack, J.M. (1997) Chemical composition and deposition of rain in the Central Amazon, Brazil. *Atmospheric Environment* 31, 207–217.

Forestry Problems in Africa 5

A.M. van Tienhoven[1], K.A. Olbrich[2] and T.F. Fameso[3]

[1]CSIR Environmentek, Pretoria, South Africa; [2]PO Box 12716, Nelspruit, South Africa; [3]Forestry Management, Evaluation and Co-ordinating Unit (FORMECU), Abuja FCT, Nigeria

Emissions from burning vegetation make the greatest contribution to the chemical composition of the atmosphere over Africa, whereas industrial emissions are responsible for isolated, local impacts in the vicinity of point sources. Slash-and-burn agriculture, rather than air pollution, poses a more immediate threat to Africa's natural forests. At present, the African vegetation surface shows little evidence of either direct effects of gaseous pollution or indirect, soil-mediated impacts. However, emissions from domestic and industrial sources will continue to increase as cities and industries develop. Long-range transport of the emission products also ensures that air pollutants are widely dispersed. Although the long-term impacts of increased emissions are currently not apparent, areas that are likely to be sensitive to soil acidification have already been identified.

1 Introduction

Africa's forests are threatened by a number of growing pressures. Land use change, slash-and-burn agriculture, urbanization and industrialization are taking a direct toll on both indigenous and commercial forests. Over the past decades, a less apparent threat to forest productivity and sustainability has emerged – that of air pollution.

Air pollution may result from both natural and human (anthropogenic) processes. Natural sources include volcanic and geothermal eruptions, vegetation fires, biogenic emissions, wind-blown soil, pollen, spores and sea spray

 Air Pollution and the Forests of Developing and Rapidly Industrializing Regions (eds J.L. Innes and A.H. Haron)

particles (Smith, 1981). A variety of combustion and industrial activities make up anthropogenic sources. Humans have long used fire to stimulate new growth of grass for grazing but it is the last century of industrial air pollution that gives the most cause for concern.

The African continent is plagued by political instability, war, famine, drought and disease. Compared with such critical issues, it is not surprising that air pollution is not recognized as a priority in any African country. As an example, only two African countries (South Africa and Kenya) have representatives on the International Union of Air Pollution Prevention and Environmental Protection Associations (IUAPPA).

2 Industrial air pollution in Africa

Environmental legislation controlling industrial air pollution in Africa is generally lax, with few or no air quality standards (Semwayo and Simukanga, 1998). Countries such as South Africa and Zambia have air quality guidelines but these are not legally enforceable. Where legislation does exist, there is little enforcement of air pollution control laws. There are no regional political agreements on transboundary air pollution. Toxic substances such as cadmium, benzene and asbestos are not controlled as they are in Europe and North America. On the contrary, many African governments ignore polluting industries in the drive to encourage economic growth. Low quality fuels, leaded fuels and diesel, coupled with poorly maintained vehicles, further exacerbate the pollution situation. Meteorological conditions are the final factor that may worsen air pollution, as sunlight and stable wind conditions encourage the build-up of ozone (McCormick, 1997).

Sulphur dioxide, nitrogen oxides, ammonium, ozone and lead are the most important industrial air pollutants on the African continent. While the air pollution control authorities in Europe are already focusing attention on volatile organic carbon compounds and heavy metals, their African counterparts are still dealing with basic contaminants such as sulphur dioxide.

The urban pollution problems experienced by the cities of Los Angeles, London and Tokyo do not occur on the same scale in Africa. The majority of the population is rural and most of the continent's fuel requirements are primarily met by firewood. Apart from Zambia, South Africa and Nigeria, there is very little heavy industry in Africa despite the world's richest mineral field extending through most of the southern African countries. Thus, air pollution, with the exception of isolated regional problems, is not an immediate concern.

However, population growth, increasing urbanization of the population, increased energy demands and increased vehicle numbers will all contribute to future air pollution problems in African cities.

2.1 Current and predicted future industrial development in Africa

Since 1973, Africa's commercial energy consumption has leapt 145%, from 89.7 to 219.2 Mt of oil equivalent (McCormick, 1997). Future industrial development will most likely focus on mineral reserves that have not yet been exploited. There are considerable reserves of nickel, petroleum and phosphate in Burundi which are likely to be mined in the next decade (Baradandikanya, personal communication, 1997, Côte d'Ivoire). South Africa is actively trying to encourage foreign investment with the promise of cheap electricity. Nigeria is also experiencing industrial growth based on its reserves of natural gas and oil.

The Stockholm Environment Institute has estimated future sulphur emissions based on predictions of industrial growth (Fig. 5.1). Areas of concern

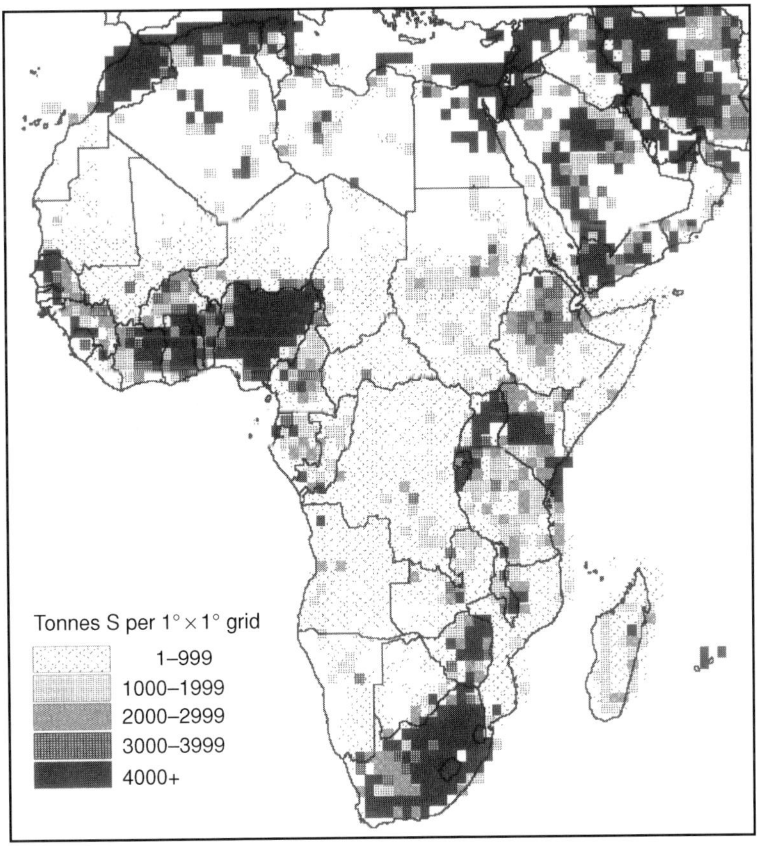

Fig. 5.1. Projected sulphur emissions in the year 2050 for the continent of Africa. (Courtesy of Stockholm Environment Institute, York.)

include West Africa, particularly Nigeria, and North African countries such as Morocco. In southern Africa, Zimbabwe and South Africa are highlighted, while high emission values are also anticipated for Kenya.

2.2 Focus on southern Africa

In 1993, the Southern African Development Community (SADC) commissioned a study on the transboundary movement of pollutants, specifically sulphur, in southern Africa (Sivertsen *et al.*, 1995). Sulphur emission estimates are available for the southern African region, which is defined as those countries south of the 18° parallel (Table 5.1).

Botswana's emissions per capita seem alarmingly high, compared with the European average of 29 kg S per capita, but the population of Botswana is small – approximately 1.6 million people occupying 582,000 km^2 of land. Large portions of the subcontinent are sparsely populated, so the average emission density of 0.3 t S km^{-2} experienced by southern African countries is much less than the average European emissions of 1.9 t S km^{-2}. However, in highly industrialized areas such as the South African highveld (Fig. 5.2), emission densities are similar to those in Europe (Sivertsen *et al.*, 1995).

Smelters processing ore from the rich mineral deposits are the major sources of air pollutants in southern Africa. Ore deposits are located along a north–south axis from eastern South Africa, through eastern Botswana and into western and northern Zimbabwe and north-eastern Zambia. Mineral deposits are also found in Botswana and Angola. Sulphur dioxide is also emitted from numerous power stations hosted by the extensive coal fields of South Africa (Sivertsen *et al.*, 1995).

On the issue of transboundary pollution the SADC study came to the following conclusions:

Table 5.1. Estimated annual emissions of sulphur in southern Africa, based on 1992 data (Sivertsen *et al.*, 1995).

Country	Total emission (t S year^{-1})	Emission density (t S km^{-2})	Emissions per capita (kg S per person)
Botswana	206,273	0.35	156
Lesotho	991	0.03	0.6
Mozambique	5,447	0.007	0.3
Namibia	30,795	0.057	22.0
South Africa	740,032	0.65	19.2
Swaziland	3,821	0.22	4.8
Zimbabwe	116,304	0.30	14.7
Total	1,103,663	0.3	15.8

Fig. 5.2. Coal-fired power station at Witbank on the South African highveld.

- Emissions from South African sources may be transported south and south-eastwards to Lesotho, Swaziland and possibly south-west Mozambique.
- Emissions from Botswana may cause impacts on the north-east parts of South Africa and south-west Zimbabwe.
- Emissions from large sources in Zimbabwe may be transported into Botswana and South Africa.
- Zambian emissions from copper smelters may be transported into Botswana and South Africa.

Where possible, specific countries are discussed in more detail below. Sivertsen *et al.* (1995) provide a good overview of sulphur emissions for the southern African countries (Botswana, Lesotho, Mozambique, Namibia, South Africa, Swaziland and Zimbabwe).

2.3 Country-specific issues

Mozambique

Mozambican sources of sulphur dioxide include the single power station in Maputo, a few industries that use domestic coal and liquid fuel, and emissions from vehicle traffic. Total emissions in 1990/91 amounted to 5447 t S annually. Although these emissions are small, the potential for them to increase is considerable. The economy of Mozambique is slowly recovering after years of war. New industrial developments, such as an aluminium smelter and an iron reduction and iron slab plant, are planned for Maputo. Rich energy sources within the country, such as the Phande gas fields, will be exploited in the future. Mozambique thus has the potential to become a net exporter of energy (Sivertsen *et al.*, 1995). Air pollution and its impacts may be insignificant at present but are likely to increase substantially in the future.

Nigeria

Particulate matter is recognized as the major contributor to atmospheric pollution in Nigeria (Akeredolu, 1989). Sources of particulate matter include re-entrainment of dust by vehicles, biomass burning, waste incineration, industrial activities and wind-blown dust from the Sahara desert (Harmattan dust). In addition, at least 77% of the total natural gas produced in Nigeria is flared routinely, emitting considerable quantities of suspended particulate matter, sulphur dioxide, nitrogen oxides and carbon monoxide. In 1994, gas flaring in the Niger Delta was estimated to release 35 Mt of CO_2 (Moffat and Linden, 1995). Gas flaring is reported to have an adverse effect on vegetation growth. A study at the Izombe Flow Station (Imo State) showed complete yield loss in crops grown 200 m from the station, 45% loss for those grown 600 m away and 10% yield loss at a distance of 1 km (NEST, 1991). The use of motor vehicles has increased dramatically, leaping 250-fold between 1953 and 1985. Estimates of the current vehicle population are not available. Available emission data for the country are summarized in Table 5.2.

The lead content of Nigerian vehicle fuel is the highest in the world – an average of 0.74 mg l^{-1}, compared with the European Union standard of 0.15 mg l^{-1} (Moffat and Linden, 1995). Levels of lead in bioindicators such as mosses, tree leaves or bark are correlated with traffic density (Akerodolu, 1989).

On the basis of emission inventories, there is a strong possibility that acidification is occurring in the main industrial areas around Lagos, Kano and Warri/Port Harcourt. Acidification resulting from industrial activities may not yet be obvious because most industrial concerns were only established in the 1980s. Incidents of low rainfall pH (pH 3.71) have already been recorded (Akeredolu, 1989). Industrial emissions have had significant negative effects at a local scale, but emissions from biomass burning are generally considered to be more important than industrial air pollution (Isichei and Akeredolu, 1988; see also Chapters 10 and 11). However, as industrialization progresses, concerns about atmospheric pollution from industry may shift from a local to a regional focus.

Table 5.2. Inventory of sources of atmospheric pollution in Nigeria (adapted from Akeredolu, 1989).

Atmospheric pollutant	Source	Estimated emission ($t\ year^{-1}$)
Particulate matter	Dust mobilization	
	unpaved roads	612,000
	paved roads	187,000
	Wind-blown (Harmattan) dust	160,000–600,000
	Domestic waste burning	58,400
	Fuelwood burning	584,000
	Bush burning (shifting cultivation practices)	871,000
	Natural gas flaring	2,700
	Primary energy production	44,700
Carbon monoxide	Fuelwood burning	6,420,000
	Natural gas flaring	5,400
	Primary energy production	73,500
Carbon dioxide	Biomass burning	No data
	Natural gas flaring	77,800,000
	Primary energy production	77,720,000
Nitrogen oxides	Fuelwood burning	40,900
	Natural gas flaring	27,000
	Primary energy production	1,170,000
Sulphur oxides	Natural gas flaring	160
	Primary energy production	5,760
Hydrocarbons	Fuelwood burning	87,600
Lead	Motor vehicle emissions	5,200

South Africa

The vast coal fields of the South African highveld supply about 96% of Africa's hard coal (85.4 Mt in 1995; ESKOM, 1996). Two-thirds of this coal is used for the generation of electricity; the remainder is either exported or used as raw material for the petrochemical industry. Over 50% of the electricity consumed in Africa is produced in South Africa (ESKOM, 1994). At the present rate of extraction, South African coal reserves are anticipated to last for another 300 years.

Most of South Africa's coal is burned on the highveld where industrial plants such as petrochemical works, smelters, manufacturing plants and power plants are to be found. These plants emit about 2 Mt of sulphur dioxide per year and so the emission densities on the highveld match those encountered in Europe – 1–20 t S km^{-2} (Sivertsen *et al.*, 1995; Wells *et al.*,

1996). Furthermore, ESKOM (the South African electricity utility) anticipates that its own annual sulphur dioxide emissions will increase from 1.2 to 1.5 Mt by 2010, when the new-generation power stations will be commissioned (C.R. Turner, personal communication, 1996, Cleveland, South Africa). Table 5.3 shows the emissions from scheduled industries on the highveld compared with the rest of South Africa, illustrating the concentrations of pollution on the highveld.

ESKOM pledged to reduce the real price of electricity by 15% so as to become the world's lowest-cost producer of electricity by the year 2000 (ESKOM, 1996). There is considerable concern that cheap electricity will encourage energy-intensive industries to invest in South Africa. While economic growth may be stimulated, the resource implications of increased electricity requirements and increased waste have been inadequately addressed. The concern is that the impacts of these industries both on the environment and on society are not reflected in the real cost of electricity (van Horen, 1996).

The accumulation of atmospheric pollutants is exacerbated by the poor dispersion climate that prevails on the highveld, particularly in the winter months. With the development of stable boundary conditions, low-level emissions are not well dispersed. Emissions that are expelled above the boundary layer can be transported away from the area but can also be recirculated by the low-level jet at both the regional and subcontinental scale (Held *et al.*, 1996).

Particulate emissions from household coal and wood burning have been identified as the major cause of poor air quality in urban areas of South Africa. The health of an estimated 20–24 million people in exposed communities is expected to be adversely impacted (Murley, 1995). Programmes to introduce low-smoke fuels to affected communities are currently under way but have met with limited success (Asamoah *et al.*, 1997).

South Africa is the 18th largest source of greenhouse gas (GHG) emissions in the world, contributing about 1.2% of the global greenhouse effect (Table 5.4). South Africa also accounts for 15% of the greenhouse gas emissions from the African continent. The most significant emissions of carbon dioxide in South Africa come from power stations (142.9 Mt CO_2 in 1994) (Scholes and van der Merwe, 1993; van Horen, 1996).

Table 5.3. Scheduled emissions at all stack heights from South Africa and from the industrial highveld region (t year^{-1}) (Wells *et al.*, 1996).

Pollutant	Total emissions in South Africa	Emissions in industrial highveld region	Highveld emissions (% of total)
Particulates	331,399	285,405	86
Sulphur dioxide	2,120,452	1,986,193	94
Nitrogen oxides	1,004,716	913,486	91

Table 5.4. South African emissions of greenhouse gases, using 1989–1991 data (adapted from Scholes and van der Merwe, 1993).

Gas	South African emissions ($Mt\ year^{-1}$)	Contribution to South African greenhouse effect (%)	South African contribution to global greenhouse effect (%)
CO_2	307.86	50.07	1.18
CH_4	2.25	23.05	0.75
N_2O	0.47	20.64	7.83
CFC	0.006407	6.24	0.64
Total		100.00	1.17

An emerging issue of concern is that of motor vehicle emissions and their anticipated increase in numbers (Annegarn, 1997). Improved land use planning, efficient public transport systems and the use of fuel-efficient vehicles will be essential if South African cities are to avoid the air pollution problems experienced by Los Angeles and Mexico City (Miller and McBride, 1998; Chapter 3).

Zambia

The copper industry of Zambia is responsible for most of the air pollution experienced in that country. Almost 60% of Zambia's urban residents live in the eight towns of the Copperbelt. Smelting of the copper ore releases an estimated 1250 t of sulphur dioxide daily (228,125–255,000 t S annually) (Perera, 1982; Semwayo and Simukanga, 1998). Besides sulphur dioxide, the air pollution mixture generated by copper smelters contains particulates, oxides of nitrogen, carbon dioxide, carbon monoxide and, to a lesser degree, heavy metals such as lead.

Zambian air quality guidelines are very strict, comparable to those of the World Health Organization. However, the ambient sulphur dioxide guidelines of Zambia, the World Bank, South Africa and the World Health Organization, as well as the European Union and US-EPA standards, are consistently exceeded in the vicinity of the smelters. Human health effects were first noted in the late 1930s, although commercial copper production only began in 1931. Acidic deposition has apparently already been recorded and reportedly affects crop growth near the smelters (McCormick, 1997). Toxic levels of sulphur dioxide are apparently responsible for the poor growth of tomato plants and also for flecks on the leaves of mango trees grown in the vicinity of the smelters (Perera, 1982). Episodes of poor visibility are common and are attributed to clouds of sulphuric acid mist or sulphur trioxide. There are also anecdotal reports of air pollution damage to townships and vegetation close to copper smelters. However, the intense extraction of trees and shrubs for fuelwood makes it difficult to ascribe vegetation damage directly to air pollution.

Concerns have been raised about visibility and about the long-term, long-distance effects of SO_2 pollution. Specifically, the freshwater fishing, agricultural and forestry industries are viewed as susceptible to environmental damage from SO_2 emissions generated by the copper smelters (Perera, 1982; McCormick, 1997).

Zimbabwe

The total sulphur emission of Zimbabwe is estimated at 116,304 t S annually. Industry and power production are responsible for 96% of these sulphur emissions. Industries include brick factories, foundries, cement factories, a sulphuric acid plant, a pulp and paper plant, a fertilizer plant, various metallurgical plants (chromium, nickel and arsenic processing) and steel manufacture (Sivertsen *et al.*, 1995).

Fuelwood supplies 98.2% of the domestic energy needs of Zimbabwe. Each rural family is estimated to consume 12.5 kg of fuelwood daily, so total fuelwood consumption amounts to 5475 kt annually. Burning of fuelwood contributes to both global warming through the emissions of greenhouse gases and to acidification through the release of sulphur and nitrogen compounds.

2.4 Air pollution issues of potential concern for Africa

The chemistry of the atmosphere in the tropics of Africa is largely determined by the extent of biomass burning (Delmas *et al.*, 1991; Menaut *et al.*, 1991). Over southern Africa, emissions of ozone precursors from the vegetated surface are considered to be greater than the contributions from industrial activities (Connors *et al.*, 1991). Table 5.5 shows the estimates of ozone-forming trace gases from biogenic, pyrogenic and industrial emissions in southern Africa. Whatever the source of the pollutants, whether from biomass burning or industrial pollution, they follow the same reaction pathways and effect the same result on the chemistry of the atmosphere.

The Southern African Fire–Atmosphere Research Initiative (SAFARI-92) found that vegetation fires in southern Africa account for a substantial amount of photochemical oxidants and haze over the subcontinent. Smoke from savannah fires produces the precursors for tropospheric ozone formation, other trace gases and aerosols that accumulate over the tropical oceans surrounding Africa. Ozone concentrations over Africa range from 60 to 90 ppb in the dry season (December–March). By comparison, ozone levels in the industrialized northern hemisphere normally range between 50 to

Table 5.5. Summary of the emissions of ozone-forming trace gases (10^{12} g year^{-1}) from broad categories of sources in Africa south of the equator (9.6×10^{12} m^2) (Scholes and Scholes, 1998).

Source category	Methane (CH_4)	Non-methane volatile organic compounds	Carbon monoxide (CO)	Nitric oxide (NO)
Pyrogenic	0.5	0.54	14.9	1.04
Biogenic	> 0.32	30–500	0.5	0.29
Industrial and domestic	> 2.59	0.61	5.6	1.95
Total	4	100?	21	3.3

60 ppb, compared with 25–30 ppb in the relatively unindustrialized southern hemisphere (Marenco *et al.*, 1990). Biomass burning is also a significant contributor to the overall increase in greenhouse gases that has occurred over the last 150 years, accounting for 10–25% of current emissions (Lindsey *et al.*, 1996).

Although the major sources that contribute to the tropospheric ozone maximum over the southern Atlantic Ocean each September–October have been identified, there is still uncertainty over the relative contributions of these sources. Consideration of the seasonal patterns and total amounts of trace gases emitted does not allow any of the major sources to be eliminated. Thus, the tropospheric ozone maximum possibly results from a combination of the tail end of pyrogenic emissions (August–October) and the beginning of biogenic emissions (September–October). Scholes and Scholes (1998) argued that if such is the case, the tropospheric ozone maximum over the southern Atlantic Ocean has probably recurred for millions of years and is thus a natural phenomenon.

Following the successful 1992 SAFARI experiment, a second 3-year field campaign started in 1999, with the broad aim of understanding the southern African biogeophysical system. The SAFARI 2000 initiative will explore linkages between land–atmosphere processes and the relationship of biogenic, pyrogenic and anthropogenic emissions, and the consequence of their deposition, on the functioning of southern African systems.

Lacaux *et al.* (1992, cited in Lacaux *et al.*, 1996) found that acidic deposition to the African equatorial forest amounted to 0.74 kg H^+ ha^{-1} year^{-1}, of which biomass burning contributed 80%. In comparison, the eastern USA, which is considerably industrialized, receives 0.67 kg H^+ ha^{-1} year^{-1}. Rainfall collected in the equatorial forests of the Congo has revealed that the effect of savannah burning is considerable, with 90% of nitrates, 80% of ammonium and 30–75% of organic acids (acetic and formic acid) originating from savannah sources. Rainfall acidity in northern Congo is thus high, with a mean pH of 4.45 (Lacaux *et al.*, 1991).

African countries are becoming more industrialized but still rely heavily on agriculture for most of their export products. For example, 50% of Kenyan exports are agricultural products, such as coffee and tea. If not carefully considered, increased industrialization and poor pollution control could lead to a situation where air pollution may threaten the agricultural livelihood of a region.

3 Forests in Africa

The expanse of the Sahara desert dominates North Africa, though there are some forested areas in the Atlas mountains of Morocco. The forests of central Africa cover an area of about 2,000,000 km^2 and are enclosed by a savannah zone of 10,000,000 km^2 (Lacaux *et al.*, 1991). The southern African landscape is dominated by vast grasslands, wooded savannah and forest–savannah mosaics. Desert regions and pockets of rainforest are also present (Connors *et al.*, 1991).

In countries such as Nigeria, deforestation is occurring at an annual rate of 4.8% to make way for shifting, slash-and-burn type agriculture. Subsistence agriculture was previously restricted to the savannah but growing populations have now been forced to encroach into transitional vegetation and forest (Akeredolu and Isichei, 1991). Forests in Burundi are also threatened by population pressure and semi-extensive farming of crops and livestock. Only small patches of high altitude and gallery forest remain in forest reserves (Baradandikanya, personal communication, 1997, Côte d'Ivoire).

Areas of land that have been deforested to make way for subsistence agriculture could potentially be afforested with commercial tree species. Limitations to afforestation plans by socio-economic factors are beyond the scope of this chapter. Possibly of greater importance would be the limitations placed by water and nutrient availability. Once the nutrient cycle between the leaves of the forest canopy and the forest litter layer has been disrupted by forest burning, it is questionable whether it can be adequately restored. Nutrient inputs through fertilizers would be a financial constraint on the success of reforestation programmes. An afforestation programme is apparently under way in Burundi in which *Eucalyptus* and *Pinus* species will be grown to meet fuelwood and construction requirements (Baradandikanya, personal communication, 1997, Côte d'Ivoire).

Community forestry programmes in South Africa also aim to provide a sustainable source of fuelwood and construction materials for rural communities. Through genetic improvement programmes, there is also considerable scope for the expansion of commercial forestry. However, afforestation in South Africa is presently limited by consideration of the national water resource. Future afforestation permits will be integrated with a water resource allocation system that is currently under development (Department of Water Affairs and Forestry, 1997).

4 Potential impacts of air pollution on African forests

4.1 Current evidence of air pollution impacts on African forests

Conditions of high solar radiation, high air temperature and the likelihood of high concentrations of photochemical oxidants occur over much of Africa. These factors increase the risk of air pollution damage to vegetation but clearcut evidence is either lacking or is species-specific. The responses of plants to air pollutants in hot, dry climates may be influenced by water and temperature stresses. Patterns of temperature-dependent plant responses to air pollutants seem quite different between species, so generalizations about the influence of exposure temperature on resistance of plants to air pollutants are limited. Similarly, water stress may protect plants from air pollutants. Stomata close in response to severe water stress to reduce the loss of water by transpiration. If the stomata are closed, the uptake of air pollutants decreases and damage to plants is reduced (Schenone, 1993).

Very little work has been done on the impacts of air pollution on African vegetation. Schenone (1993) cites some studies on tropical plants in India and Australia. O'Connor *et al.* (1974, cited in Schenone, 1993) determined the relative susceptibility of 141 Australian tree and shrub species to acute SO_2 injury. The genera *Acacia* and *Eucalyptus* were found to be the most sensitive, suffering acute leaf injury after exposure to 1000 nl SO_2 l^{-1} for 3 h. Such high concentrations of sulphur dioxide are unlikely to occur naturally. Nevertheless, the *Acacia* genus is widespread in Africa, and several *Eucalyptus* species are important commercial timber plants. The response of these species to long-term exposures to elevated sulphur dioxide may not be immediate.

In Zambia, trees about 2 km downwind of the Nkana copper smelter are reported to be damaged. However, no baseline exists so the evidence is inconclusive (Semwayo and Simukanga, 1998).

South African forests

Commercial forestry in the eastern and south-eastern parts of the Mpumalanga province of South Africa is at possible risk from air pollution. The major source of air pollutants is from the industrial highveld region about 100–200 km west of the forestry region. Prevailing winds transport air pollutants from the highveld to the commercial forests that are planted in high rainfall areas. These forests potentially receive high loads of atmospheric deposition.

Mist events are common in much of the forestry area. Olbrich (1993) found that the volume of water received as mist was equivalent to the volume of rainfall over the same monitoring period. The mist samples also had higher concentrations of chemicals – particularly sulphate, which was double that found in rainfall.

Olbrich (1990, 1995) investigated foliar symptoms that were apparently caused by air pollution on needles of *Pinus patula* Schlecht et Cham. The symptoms included chlorotic mottling, flecking, broad and narrow chlorotic banding and tip burn. Most trees exhibited some type of chlorosis, irrespective of their proximity to pollution sources. Only one symptom, namely broad banding, appeared to be distributed in relation to the potential pollution impact at the site. Site-, tree- or soil-related factors could not account for the observed distribution of the broad banding. Although air pollution is the prime suspect, it is still debated whether these symptoms are caused by air pollution, by other stresses or by a combination of these.

A number of studies have focused attention on the indirect effects of air pollution on trees through changes in soil chemistry. Soil chemical characteristics such as soil pH in water, Ca : Al ratios, concentration of exchangeable bases, percentage clay content and acid-neutralizing capacity have been used to assess soil sensitivity to acidic deposition (Olbrich and Du Toit, 1993; Olbrich, 1995). Soil sensitivity in Mpumalanga has also been mapped using the critical loads approach. Factors such as geology, land use and soil type were used to define soil sensitivity classes (Olbrich *et al.*, 1995). In general, the areas most sensitive to acidic deposition corresponded to afforested regions with highly weathered soils and low acid-neutralizing status. Fey *et al.* (1995) assessed soil sensitivity to acidification using a 'quick test' of acid-neutralizing capacity that gave comparable results to methods developed in Europe for assessing soil sensitivity.

There is emerging evidence to suggest that commercial forests of the Mpumalanga escarpment may be saturated with nitrogen. Symptoms include the lack of growth response to the addition of N fertilizers, low C : N ratios of the litter layer, high total N levels in the foliage and leaching of N compounds from the soil (Jacobs, unpublished observations). Streams draining forested catchments also show higher levels of nitrate than those draining adjacent grassland catchments (Nowicki, 1997). Whether these responses are a consequence of afforestation or enhanced interception by the tree canopy, or a combination of both, is still unclear.

The indirect impacts of acidic deposition on soil fertility, nutrient cycling processes and tree nutrition are particularly important in South Africa because of the short harvest rotations (6–12 years for eucalypt species and 15–30 years for pines; ISAF, 1993). Frequent harvesting results in the export of base cations from the site. Unless base cations are replaced through mineral weathering, fertilization or atmospheric deposition, the site may ultimately become acidified.

Carlson (1992) investigated the effects of artificially acidified precipitation on cation uptake by *Pinus patula. P. patula* and *Eucalyptus grandis* Hill ex Maiden are the two main commercial forest species grown in South Africa. Carlson concluded that acid rain had the potential to change the uptake of Ca and Mg cations by decreasing the availability of these cations in the soil, alter the mycorrhizal composition of the roots and increase the trivalent aluminium cation concentration in the soil solution.

4.2 Potential risk of pollution impacts on African forests

The Stockholm Environment Institute (SEI) has assessed global ecosystem sensitivity to acidic deposition using a reclassification of the FAO soil map of the world (FAO, 1995), based on soil chemistry data obtained from the International Soil Reference and Information Centre (ISRIC) (Batjes, 1995). Large areas of central tropical Africa are considered to be highly sensitive to acidic deposition and cover both forest and non-forest ecosystems (Fig. 5.3). Sensitive areas in Africa are quite sparsely populated and have low emissions of sulphur and nitrogen, except for South Africa, Zambia and Nigeria. Agricultural lands in marginal areas are more susceptible to acidic deposition as these lands are not limed as they would be in more-developed countries. There is a large amount of uncertainty surrounding the sensitivity assessment; nevertheless the map has been subject to scrutiny by regional experts and is generally deemed to be valid (Kuylenstierna *et al.*, 1995; Cinderby *et al.*, 1998).

Lacaux *et al.* (1991) have already shown that large inputs of acidity to equatorial forests originate from biomass burning. Vegetation fires are a natural phenomenon on the African continent, and have been so for millennia (Goldammer, 1991). The question of whether the additional acid inputs from industrial emissions will impact severely on African ecosystems remains inconclusively answered.

Many tropical soils are acidic and highly weathered and have high concentrations of iron and aluminium; thus they would not be expected to show large changes in acidity as a result of increased acidic deposition (McDowell, 1988). Data from P.G.H. Frost (personal communication, 1997, Institute of Environmental Studies) show that the dominant species of miombo woodland, namely msasa (*Brachystegia spiciformis*), which occurs over much of Central Africa, can be found over a broad range of soil pH values. Changes in soil pH are thus unlikely to impact severely on *Brachystegia*. Whether the same can be said of other tropical species requires further study.

Similarly, tropical rainforest has a relatively high rate of leaf turnover, and already occurs on naturally acidic soils that are saturated with aluminium. These forests probably avoid soil acidification problems by conducting their nutrient cycling in a dense root and mycelium network that is above the mineral soil (R.J. Scholes, personal communication, 1997, Pretoria).

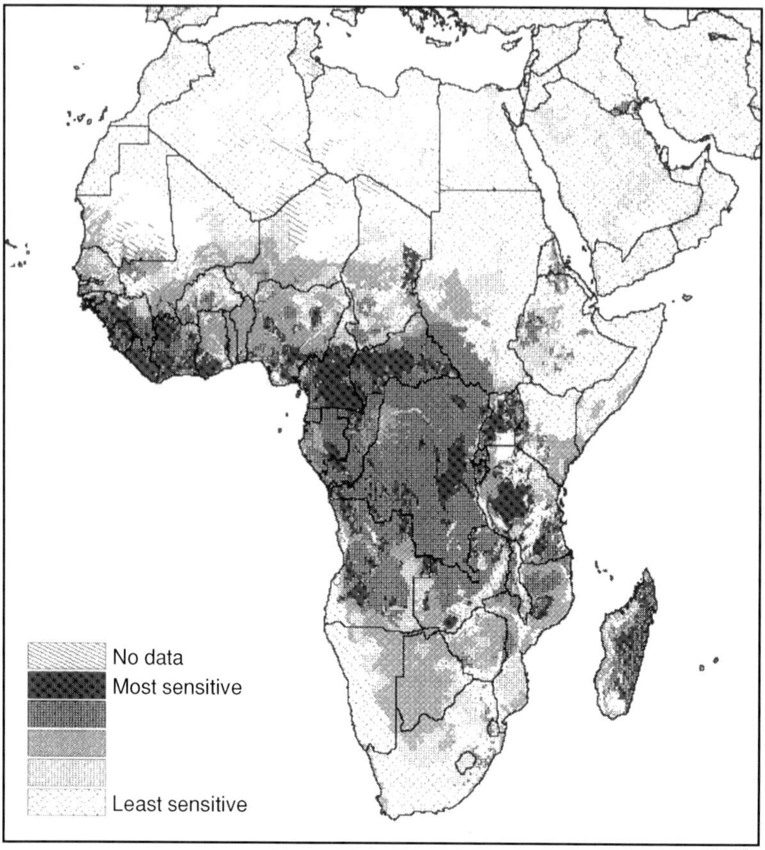

Fig. 5.3. African ecosystem sensitivity to acidic deposition. (Courtesy of Stockholm Environment Institute, York.)

5 Summary and conclusions

Considered on a continental scale, industrial emissions contribute a minor portion of the atmospheric pollution of Africa. Biomass burning is generally recognized to be the major factor affecting atmospheric chemistry, whereas industries are only considered to cause isolated, local environmental damage. Impacts of air pollution at a regional scale are not clearly evident. Natural forests are more immediately threatened by slash-and-burn agriculture than by the effects of industrial air pollution. The African continent is, however, experiencing high population growth rates, increasing levels of industrialization, and increasing urbanization – factors that will all conspire to increase levels of atmospheric pollution from both industrial and domestic sources.

Fig. 5.4. Plantations, such as this stand of *Pinus elliottii*, could be adversely affected by air pollution from the highveld.

At present, neither the direct effects of gaseous pollutants nor the indirect effects of soil-mediated impacts are apparent on the African vegetation surface. Areas that are likely to be sensitive to acidification have already been identified on a preliminary basis but further research is needed to assess the acidification potential of African soils, and to quantify their mineral weathering rates. We need to understand plant responses to atmospheric pollutants under realistic pollutant mixes and concentrations – particularly for commercial forestry species. The influence of the climatic conditions under which exposure to pollutants occurs is also an area requiring further investigation.

Over and above the scientific research priorities, other recommendations include the development of appropriate policies and legislation for air pollution control. Regional collaboration between countries is essential, particularly on the issue of transboundary air pollution and the possible effects on neighbouring nations.

It is reassuring to note the growing awareness of the risks from atmospheric pollution in Africa. Africa is fortunate to have the opportunity to learn from the experiences of Europe and North America and, in so doing, prevent the large-scale environmental damage caused by industrial emissions.

Acknowledgements

Grateful thanks are given to W.K. Hicks for comments on an earlier draft of the manuscript.

References

Akeredolu, F. (1989) Atmospheric environment problems in Nigeria – an overview. *Atmospheric Environment* 23, 783–792.

Akeredolu, F. and Isichei, A.O. (1991) Emissions of carbon, nitrogen and sulphur from biomass burning in Nigeria. In: Levine, J.S. (ed.) *Global Biomass Burning. Atmospheric, Climatic and Biospheric Implications*. MIT Press, Cambridge, Massachusetts, pp. 162–166.

Annegarn, H.J. (1997) A bird's eye view of air pollution in South Africa. In: *Proceedings of the 28th National Association for Clean Air Conference, Vanderbijlpark, Vaal Triangle, 20 November 1997*. National Association for Clean Air, Pretoria, pp. 1–13.

Asamoah, J.K., Lloyd, P., Hoets, P. and Grobbelaar, C.J. (1997) Low smoke fuel programme: preliminary results of the macro-scale experiment. In: *Proceedings of the 28th National Association for Clean Air Conference, Vanderbijlpark, Vaal Triangle, 20 November 1997*. National Association for Clean Air, Pretoria, pp. 68–75.

Batjes, N.H. (ed.) (1995) *A Homogenized Soil Data File for Global Environmental Research: a Subset of FAO, ISRIC and NRCS Profiles (version 1.0)*. Working paper 95/10. International Soil Reference and Information Centre (ISRC), Wageningen, The Netherlands.

Carlson, C.A. (1992) *Effects of Acidic Precipitation on Calcium and Magnesium Uptake by* Pinus patula. CSIR Report FOR-DEA 513. Division of Forest Science and Technology, Pretoria, South Africa.

Cinderby, S., Cambridge, H.M., Herrera, R., Hicks, W.K., Kuylenstierna, J.C.I., Murray, F. and Olbrich, K.A. (1998) *Global Assessment of Ecosystem Sensitivity to Acidic Deposition – Version 2, April 1998*. Stockholm Environment Institute, York, UK.

Connors, V.S., Cahoon, D.R., Reichle, H.G., Brunke, E.-G., Garstang, M., Seiler, W. and Scheel, H.E. (1991) Savannah burning and convective mixing in southern Africa: implications for CO emissions and transport. In: Levine, J.S. (ed.) *Global Biomass Burning. Atmospheric, Climatic and Biospheric Implications*. MIT Press, Cambridge, Massachusetts, pp. 147–154.

Delmas, R.A., Loudjani, P., Podaire, A. and Menaut, J.-C. (1991) Biomass burning in Africa: an assessment of annually burned biomass. In: Levine, J.S. (ed.) *Global Biomass Burning. Atmospheric, Climatic and Biospheric Implications*. MIT Press, Cambridge, Massachusetts, pp. 126–132.

Department of Water Affairs and Forestry (1997) *South Africa's National Forestry Action Programme*. Department of Water Affairs and Forestry, Pretoria.

ESKOM (1994) *Annual Report 1994 – Power for the People*. Corporate Communication Services, Megawatt Park, Sandton, South Africa.

ESKOM (1996) *Annual Report 1996 – Making the Most of Our Energies*. Corporate Communication Services, Megawatt Park, Sandton, South Africa.

FAO (1995) *FAO Digital Soil Map of the World*. Food and Agriculture Organisation, Rome.

Fey, M.V., Nowicki, T.E. and Dodds, H.A. (1995) *Sensitivity of the Soil Environment to the Deposition of Acidifying Air Pollutants in South Africa*. ESKOM Technology Research Report TRR/S95/201/bs. ESKOM, Johannesburg.

Goldammer, J.G. (1991) Tropical wild-land fires and global changes: prehistoric evidence, present fire regimes and future trends. In: Levine, J.S. (ed.) *Global Biomass*

Burning. Atmospheric, Climatic and Biospheric Implications. MIT Press, Cambridge, Massachusetts, pp. 83–91.

Held, G., Scheifinger, H., Snyman, G.M., Tosen, G.R. and Zunckel, M. (1996) The climatology and meteorology of the highveld. In: Held, G., Gore, B.J., Surridge, A.D., Tosen, G.R., Turner, C.R. and Walmsley, R.D. (eds) *Air Pollution and its Impacts on the South African Highveld.* Environmental Scientific Association, Cleveland, South Africa, pp. 60–71.

ISAF (1993) *South African Forestry Handbook.* Institute of Southern African Forestry, Pretoria, South Africa.

Isichei, A.O. and Akeredolu, F. (1988) Acidification potential in the Nigerian environment. In: Rodhe, H. and Herrera, R. (eds) *Acidification in Tropical Countries.* SCOPE Series 36. John Wiley & Sons, Chichester, UK, pp. 297–316.

Kuylenstierna, J.C.I., Cambridge, H., Cinderby, S. and Chadwick, M.J. (1995) Terrestrial ecosystem sensitivity to acidic deposition in developing countries. *Water, Air and Soil Pollution* 85, 2319–2324.

Lacaux, J.P., Delmas, R.A., Cros, B., Lefeivre, B. and Andreae, M.O. (1991) Influence of biomass burning emissions on precipitation chemistry in the equatorial forests of Africa. In: Levine, J.S. (ed.) *Global Biomass Burning. Atmospheric, Climatic and Biospheric Implications.* MIT Press, Cambridge, Massachusetts, pp. 167–173.

Lacaux, J.P., Delmas, R., Kouadio, G., Cros, B. and Andreae, M.O. (1992) Precipitation chemistry in the Mayombe Forest of equatorial Africa. *Journal of Geophysical Research* 97, 6195–6206.

Lacaux, J.P., Delmas, R., Jambert, C. and Kuhlbusch, T.A.J. (1996) NO_x emissions from African savanna fires. *Journal of Geophysical Research* 101(D19), 23585–23595.

Lindsey, J.A., Andreae, M.O., Goldammer, J.G., Harris, G., Annegarn, H.J., Garstang, M., Scholes, R.J. and van Wilgen, B.J. (1996) International Geosphere–Biosphere Programme/International Global Atmospheric Chemistry SAFARI-92 field experiment: background and overview. *Journal of Geophysical Research* 101(D19), 23521–23530.

Marenco, A., Medale, J.C. and Pricur, S. (1990) Study of tropospheric ozone in the tropical belt (Africa, America) from STRATOZ and TROPOZ campaigns. *Atmospheric Environment* 24A, 2823–2834.

McCormick, J. (1997) *Acid Earth – The Politics of Acid Pollution*, 3rd edn. Earthscan Publications, London.

McDowell, W.H. (1988) Potential effects of acid deposition on tropical terrestrial ecosystems. In: Rodhe, H. and Herrera, R. (eds) *Acidification in Tropical Countries.* SCOPE Series 36. John Wiley & Sons, Chichester, pp. 117–135.

Menaut, J.-C., Abbadie L., Lavenu, F., Loudjani, P. and Podaire, A. (1991) Biomass burning in West African savannahs. In: Levine, J.S. (ed.) *Global Biomass Burning. Atmospheric, Climatic and Biospheric Implications.* MIT Press, Cambridge, Massachusetts, pp. 133–142.

Miller, P.R. and McBride, J.R. (eds) (1998) *Oxidant Air Pollution Impacts in the Montane Forests of Southern California. A Case Study of the San Bernardino Mountains.* Springer, New York.

Moffat, D. and Linden, O. (1995) Perception and reality: assessing priorities for sustainable development in the Niger River delta. *Ambio* 24, 7–8.

Murley, L. (ed.) (1995) *Clean Air Around the World – National Approaches to Air Pollution Control*, 3rd edn. International Union of Air Pollution Prevention and Environmental Protection Associations, Brighton, UK.

NEST (Nigerian Environmental Study/Action Team) (1991) Nigeria's threatened environment: a national profile. *NEST Newsletter*, pp. 124–129.

Nowicki, T.E. (1997) The impact of plantations of *Pinus* spp. on the chemical properties of soils and stream waters in South African upland catchments. PhD thesis, Department of Geological Sciences, University of Cape Town, South Africa.

O'Connor, J.A., Parbery, D.G. and Strauss, W. (1974) The effects of phytotoxic gases on native Australian plant species. Part 1. Acute effects of sulphur dioxide. *Environmental Pollution* 7, 7–23.

Olbrich, K.A. (1990) *The Geographic Distribution of Possible Air Pollution Induced Symptoms in the Needles of* Pinus patula. CSIR Report FOR-1. Division of Forest Science and Technology, Pretoria, South Africa.

Olbrich, K.A. (1993) *The Chemistry and Frequency of Mist Events on the Eastern Transvaal Escarpment*. Report to the Department of Water Affairs and Forestry, CSIR Report FOR-DEA 657. Division of Forest Science and Technology, Pretoria, South Africa.

Olbrich, K.A. (1995) *Air Pollution Impacts on South African Commercial Forests: a Research Synthesis: 1990–1995 and Strategy for the Future*. CSIR Report FOR-DEA 874. Division of Forest Science and Technology, Pretoria, South Africa.

Olbrich, K.A. and Du Toit, B. (1993) *Assessing the Risks Posed by Air Pollution to Forestry in the Eastern Transvaal, South Africa*. Confidential report to ESKOM; CSIR report FOR-C 214. Division of Forest Science and Technology, Pretoria, South Africa.

Olbrich, K.A., van Tienhoven, A.M., Skoroszewski, R., Zunckel, M. and Taljaard, J. (1995) *A Prototype Acid Deposition Risk Advisory System for the Eastern Transvaal Province*. CSIR Report FOR-I 547. Division of Forest Science and Technology, Pretoria, South Africa.

Perera, N.P. (1982) Environmental problems of copper mining and refining in Zambia. *UNEP Industry and Environment* (January/February/March edition), pp. 10–13.

Schenone, G. (1993) Impact of air pollutants on plants in hot, dry climates. In: Jackson, M.B. and Black, C.R. (eds) *Interacting Stresses on Plants in a Changing Climate*. NATO ASI Series, Vol. 16. Springer-Verlag, Berlin, pp. 139–152.

Scholes, R.J. and Scholes, M.C. (1998) Natural and human-related sources of ozone-forming trace gases in southern Africa. *South African Journal of Science* 94, 422–425.

Scholes, R.J. and van der Merwe, M.R. (1993) *Greenhouse Gas Emissions from South Africa*. CSIR Report FOR-DEA 607. Division of Forest Science and Technology, Pretoria, South Africa.

Semwayo, D. and Simukanga, S. (1998) Perspectives on atmospheric pollution issues in (southern) Africa. In: Kuylenstierna, J.C.I. and Hicks, W.K. (eds) *Regional Air Pollution in Developing Countries – Background Document for Policy Dialogue, Harare, September 1998*. Stockholm Environment Institute, York, UK, pp. 109–113.

Sivertsen, B., Matale C. and Pereira, L.M.R. (1995) *Sulphur Emissions and Transfrontier Air Pollution in Southern Africa*. Report No. 35. SADC Environment and Land Management Sector Coordination Unit, Pretoria, South Africa.

Smith, W.H. (1981) *Air Pollution and Forests: Interactions Between Air Contaminants and Forest Ecosystems*. Springer Series on Environmental Management. Springer-Verlag, New York.

Van Horen, C. (1996) *Counting the Social Costs – Electricity and Externalities in South Africa. Industrial Strategy Project.* Élan Press, EDRC, University of Cape Town, South Africa.

Wells, R.B., Lloyd, S.M. and Turner, C.R. (1996) National air pollution source inventory. In: Held, G., Gore, B.J., Surridge, A.D., Tosen, G.R., Turner, C.R. and Walmsley, R.D. (eds) *Air Pollution and its Impacts on the South African Highveld.* Environmental Scientific Association, Cleveland, South Africa, pp. 3–9.

Forestry Problems Related to Air Pollution in Central Asia

6

D.I. Odintsov

Federal Forest Service of Russia, Moscow, Russia

The very large area of forest lands within the territory of Russia creates a number of unique problems for the efficient management of the resource. In the eastern part of the country, pollution problems exist around point sources such as Noril'sk and Bratsk, but the low population to the east and the relatively low levels of industrialization mean that pollution is not a major issue away from population centres. Instead, fire presents a major challenge, and the emissions from these fires are of major national and international importance. New strategies are being adopted that will help with the location and extinguishing of fires, but the issue remains an important challenge.

1 Introduction

About 25% of the forests of the world are located in Russia, covering a territory of about 1.2 billion ha (Kukuev, 1997). Most of these forests (*c.* 73%) lie in the Asian part of the Russian territory. Of the 932.8 Mha of forest in the Asian part of Russia, only 53% are covered by inventory data obtained by land assessment and interpretation methods, and the data for a number of forest management units (*leskhoz*) – Yakutia, Tuva, Amur, Chita, Irkutsk and Khanty-Mansi – are obsolete. Information about the remaining forests is obtained through assessment of reserved forests (268.2 Mha) and remote sensing studies undertaken in the 1950s (172.1 Mha). The forests of the Asian part of Russia are dominated by larch (*Larix* spp.), although a number of broadleaved species, including Mongolian oak (*Quercus mongolica*) and stone birches (*Betula* spp.), are also present, accounting for about 82 Mha. Most of the forest resource (91% of the total growing stock) is under the direct control of the Federal Forest Service of Russia.

Russia borders the countries of Central Asia, and contributes to the transboundary transfer of polluting substances. Theoretically, Russia could be a possible contributor to some of the environmental changes in these countries caused by the deposition of polluting substances from the atmosphere, assuming that seasonal air currents result in emissions from Russian sources being carried to those countries. Consequently, the study of atmospheric air pollution problems in Russia resulting from industry-related emissions is of particular interest to the countries of Central Asia.

Bezuglaya and Berland (1983) have indicated that the regions of Russia closest to Central Asia are primarily characterized by high pollution potentials (HPP) of between 3.0 and 4.0. The HPP is a correlation between average concentrations of equal emissions in specific and conventional regions differing in climatic conditions of the pollutant distribution. However, very little is known about ambient concentrations of pollutants in the region, and there has been no systematic data collection programme to assess the extent of transboundary transfer of pollutants from Russia to the Central Asian countries.

2 Legal basis for forest protection in Russia

In Russia, in accordance with the federal law 'On Ecological Expertise' adopted in 1996, a state ecological expertise service has been established. The year 1997 was a very important one for Russian forestry and the development of forest management there. On 22 January 1997 the State Duma (the Lower Chamber of the Russian Parliament) adopted the Forest Code of the Russian Federation. The Code was signed by the President of the Russian Federation on 29 January 1997. As part of the Forest Code, a number of legislative standards and acts were developed and approved. Of these, 16 documents were submitted for consideration by the Government of the Russian Federation and, by 1 January 1998, seven of the acts considered by the Government had been enacted. Also in 1997, the Government approved a comprehensive programme 'Forests of Russia' for the 1997–2000 period (Decree of 26 September 1997, No. 1240), with the purpose of improving forest management, forest protection and forest regeneration, and in order to ensure the rational use of forest resources. The programme concentrates on ecology, forest management and socio-economic problems.

There is increasing recognition of the important ecological role of the forests of Russia. The Government has issued a number of orders relating to the safeguarding of forests against fires. These orders have assisted in the solution of the complex problem of wildfire, as a result of the joint efforts taken by

Gosleshoz (the Federal State Forestry Department) and other federal bodies of state power and local authorities in the administrative territorial and national subdivisions of Russia.

In accordance with the Federal Programme 'Forests of Russia', Gosleshoz planned silvicultural activities for 1998 over an area of 32.5 Mha, an increase of 5% on the silvicultural work undertaken in 1997. Gosleshoz is a specially authorized agency for nature protection activities and one of its top priorities in environmental policy is the formation and development of a National Parks network. Today, under the Federal State Forestry Department, there are 33 National Parks with a total area of over 6.6 Mha. The current trend in policy is towards the expansion and improvement of the whole system of specially protected federal forest areas, with priority being given to National Parks.

Consistent with the federal law 'On Specially Protected Nature Areas' and the Decree of 10 October 1996 'On Procedure for Conduct of the State Cadastre of Specially Protected Nature Preserve Areas', specific measures were taken for the organization of the State Cadastre for all specially protected areas located on the lands of Goslesfund (the State Forest Fund) as of 1 January 1998. To date, over 2000 nature monuments are under the protection of *leshozes* (forest management units), covering an area of 0.9 Mha, and a further 800 preserves cover an area of 5.8 Mha.

The Decree of 20 May 1997, No. 611 'On State Forest Inventory of Forest Fund', resulted in the initiation in 1998 of a regular state forest inventory of the state-owned forest resources as of 1 January 1998. The inventory aims to ensure the rational use, safeguarding and protection of forests, and to introduce systematic monitoring and control over quantitative and qualitative changes occurring in Goslesfund forests. The data obtained by the inventory are submitted to the state authorities at both federal and regional levels. As of 1999, documents relating to the inventory of Goslesfund forests are updated annually.

3 Emission of air pollutants in Russia

In recent years, the impact of Russia's economic activity on the environment has been reduced, although this has primarily arisen because of the decline in industrial production rather than from any deliberate attempts to reduce pollutant emissions. At the same time, there has been a decrease in capital investment in nature conservation activities. Consequently, a number of recent ecological forecasts have predicted a worsening of environmental quality in Russia.

Recent trends in the emissions of pollutants from different segments of industry in Russia are presented in Table 6.1. The heaviest atmospheric pollution (by volume of emissions) resulted from the activity of energy-generating plants and ferrous and non-ferrous metal industries; these sectors were responsible for more than 60% of the total emissions from Russian industry.

A breakdown of the different pollutants being emitted in Russia is given in Table 6.2. Based on the data presented in Tables 6.1 and 6.2, a number of conclusions can be drawn:

- the total volume of harmful substances released into the atmosphere decreased between 1993 and 1996, and in 1996 amounted to 91.8% of the 1995 level.
- There was a decrease in emissions of many of the substances on record, with 25 of the 34 harmful substances showing a decrease. However, emissions of cadmium oxide, manganese compounds, cupric oxide, mercury, hydrogen cyanide, sulphuric acid, carbon disulphide and hydrogen fluoride all increased.

Relatively little information is available on levels of radionuclide contamination in the Asian part of Russia. Nuclear weapon testing prior to the ban on ground-level and airborne nuclear testing caused the radioactive contamination of *c.* 272,000 ha of forest in the Altai territory and the Republic of

Table 6.1. Trends in the emission of atmospheric pollutants (Mt) from stationary sources, organized by industrial sector. (Sources: Anon., 1996, 1997; Shekhovtsev *et al.*, 1997.)

Sector	1991	1992	1993	1994	1995	1996
Russian Federation	31,801.0	28,127.0	24,788.3	21,929.0	21,269.6	20,274.1
Industry	28,544.0	25,237.0	22,167.7	19,528.3	18,140.4	16,661.0
Energy production	7,570.7	6,644.8	5,898.2	5,267.4	5,017.7	4,748.5
Non-ferrous metal industry	5,088.5	4,647.5	3,795.0	3,502.0	3,693.2	3,598.1
Ferrous metal industry	4,036.5	3,571.5	3,227.1	2,730.2	2,735.3	2,535.5
Oil-refining industry	2,345.8	2,137.5	1,862.7	1,687.3	1,409.1	1,309.7
Oil-processing industry	1,436.0	1,359.7	1,190.8	1,004.2	908.6	849.1
Engineering industry	1,917.6	1,594.0	1,289.8	945.4	725.6	602.5
Gas industry	1,194.6	1,036.8	879.8	662.8	707.7	541.8
Construction materials industry	1,763.4	1,386.1	1,064.1	771.9	674.2	528.0
Coal-mining industry	236.0	268.8	384.2	686.7	626.5	595.8
Chemical and petro-chemical industry	1,182.5	1,000.0	728.8	548.6	525.0	454.1
Timber-processing and paper industry	855.3	751.0	638.1	523.3	522.2	434.3
Food industry	462.7	448.0	419.2	338.4	300.3	250.2
Defence industry	No data	No data	274.7	161.2	138.8	95.5
Light industry	170.1	151.4	129.9	95.4	74.2	64.4

Table 6.2. Trends in the emissions of atmospheric pollutants from stationary sources, arranged by substance. (Source: Shekhovtsev *et al.*, 1997.)

Indicator	1993	1994	1995
Total emissions (thousand tons)			
Solid substances	22,167.7	19,528.3	18,141.4
Liquid and gaseous substances	4,299.7	3,481.0	3,192.2
Sulphur dioxide	17,868.0	16,047.3	14,948.2
Carbon monoxide	6,952.4	6,260.6	6,168.1
Nitrogen oxides	5,162.1	4,409.7	4,201.8
Hydrocarbons (without VOCs)	2,142.8	1,839.0	1,723.5
Volatile organic carbons (VOCs)	1,735.8	2,021.5	1,751.1
	1,502.7	1,155.6	898.6
Specific pollutants (tons)			
Vanadium pentoxide	7,921.1	3,150.5	2,670.5
Cadmium oxide	6.1	5.0	18.7
Manganese and its compounds	1,333.1	922.0	987.7
Cupric oxide (calculated as copper)	6,003.5	4,940.8	5,020.1
Metallic nickel	3,764.4	3,552.9	3,322.8
Metallic mercury	7.8	6.6	7.0
Lead and its compounds	1,068.1	869.7	610.8
Chromium hexavalent	320.3	185.5	144.8
Nitric acid	762.2	493.9	430.6
Ammonia	46,803.0	39,106.5	3,6715.1
Hydrogen cyanide	489.5	2,745.2	2,060.2
Sulphuric acid	33,637.3	31,972.8	32,366.6
Arsenic, inorganic compounds	978.7	1,045.5	749.2
Soot	78,983.6	63,522.0	57,133.1
Hydrogen sulphide	18,608.9	13,506.8	12,155.3
Carbon disulphide	17,300.8	10,984.4	11,505.1
Fluoric gaseous compounds	12,427.7	11,591.4	12,139.6
Chlorine	1,382.0	1,079.1	1,062.3
Benzol	15,924.4	12,492.8	11,409.9
Xylene	35,163.2	26,410.4	21,549.3
Styrene (= vinyl benzene)	1,810.1	1,319.1	1,121.3
Toluene	42,775.4	31,709.9	28,182.3
Benzo(a)pyrene	108.6	42.5	21.5
Dichloroethane	4,264.6	3,639.3	3,240.5
Carbon tetravalent	495.7	292.9	533.9
Isopropyl alcohol	3,125.2	1,813.9	1,304.9
Phenol	2,967.3	2,401.3	2,190.1
Butyl acetate	9,014.0	5,859.3	4,568.9
Ethyl acetate	10,029.9	6,742.8	5,177.6
Formaldehyde	3,333.3	2,383.0	2,099.1
Acetone	23,132.8	14,173.1	10,759.9
Methyl mercaptan	1,485.7	959.5	883.3
Protein dust (CAV, complex albumin-vitamin)	48.5	36.0	25.1
Petroleum	462,126.1	447,200.2	346,259.3
Captured and decontaminated (%)	79.2	78.2	79.4

Altai. Problems also exist around Mayak in the southern Ural region, where an area of 700,000 ha has been contaminated as a result of a number of accidents (Kharuk *et al.*, 1996). In the whole Russian territory, the total forest area contaminated by long-lived radionuclides is in excess of 3 Mha (Strakhov *et al.*, 1997).

4 Impacts of air pollutants on Russian forests

The adverse impacts of industry on natural ecosystems is one of the major anthropogenic factors affecting the country as a whole and forests in particular. Gaseous mixtures of hydrogen fluoride, chlorine and sulphur dioxide, representing a considerable proportion of the total annual emissions in Russia, are particularly hazardous to forests. Estimates of the total area of forest in Russia that is adversely affected by pollution differ, but may be of the order of 1–1.5 Mha (Martynuk and Kasimov, 1993; Vassilieva *et al.*, 1999). The most pessimistic estimates are those of Nilsson and Shvidenko (1998), who claim that 210 Mha are at risk from sulphur deposition in the Asian part of Russia.

Pollution can affect forests by increasing the susceptibility to unfavourable climatic factors, pests and diseases, seasonal changes in the plants' growth and by decreasing species diversity. The rapid increase in the amounts of pollutants in the atmosphere has resulted in the displacement of many biogeochemical cycles (see for example Schlesinger, 1997). The adaptability and resilience of ecological systems are reduced. In extreme situations, industrial deserts void of any vegetation replace natural ecological systems (Fig. 6.1) (Kharuk, 1999; Rigina and Kozlov, 1999). Extended exposure to high pollution loads results in severe damage to forests. In most cases, the damage is readily visible, a situation very different from that found in areas of chronic pollution damage, where the damage may only be evident in the form of relatively subtle ecosystem changes.

Unfortunately, there is no uniform approach to either the assessment of forest condition or the specification of zones in industrial desert areas, and these shortcomings have impeded the collection and presentation of data for Russia. In addition, the absence of consistent data has hindered the planning of forest management activities in affected areas. The most detailed information on the impacts of industrial pollution on forests relates to 1990, when a special statistical investigation was undertaken (Anon., 1990a). Over the last 9 years, there have been only been slight changes in the situation, brought about by the decline in industrial production described above. The total area of forest that has been damaged to varying degrees or destroyed amounts to 832,461 ha (Anon., 1990a). Between 1991 and 1996, official statistics reveal that the affected area increased by another 66,000 ha (Anon., 1994, 1996, 1997).

Fig. 6.1. Industrial deserts such as this one at Nikel in north-west Russia are found in several parts of the country.

The largest forest area damaged through industrial emissions (600,000 ha) is located north of the Polar Circle in the south-eastern part of the Taimyr Peninsula, within the zone of the Noril'sk Mining and Metallurgical Integrated Enterprise's activity. This enterprise is the most important source of pollution in Russia and in Krasnoyarsky Krai, with a total volume of harmful substances emitted into the atmosphere amounting to 2,041,200 t. Of this, 95% comprises sulphur dioxide. As a result of the long-lasting pollution impacts on the territory of Taimyr Autonomous District, 13 Mha of lands formerly used for hunting and pasture are now unusable. Even in the territory of Evenkia Autonomous District, which is considered to be the least populated in Russia and thus the least industrially developed, the soil cover in some places contains abnormal concentrations of ferrous and non-ferrous metals as a result of pollution from the Noril'sk Mining and Metallurgical Integrated Enterprise. The concentrations of several heavy metals and radioactive elements (uranium) are well above the admissable concentration levels (ACLs), with ACLs for zinc being exceeded by a factor of 7–12, chromium by a factor of 12, copper by a factor of 10–260 and lead by a factor of 3–6 (Anon., 1996). More information about the impact of pollution on forests in the Noril'sk area is provided in Kharuk (1999).

Modelling has suggested that pollution is responsible for an overall decrease in forest productivity in Russia of 7.6% (Shvidenko *et al.*, 1997). In some regions, the growth decrease may be greater. However, considerable care needs to be taken over the interpretation of such models: similar growth reductions were projected for Europe (e.g. Nilsson, 1987), but there, forest growth is actually increasing (Spiecker *et al.*, 1996). The greatest danger for forests (in terms of emission amounts) is from the emissions from non-ferrous metal and chemical industries and from power plants. These release phytotoxic combinations of fluorine, sulphur, nitrogen and chlorine into the atmosphere. Coniferous stands of pine and spruce appear to be the most sensitive to these pollutants.

Under conditions of high atmospheric pollution, forest management may fail to ensure the sustainability of the forests. Consequently, there is a need to reduce emissions below the critical levels for forests. This approach should become a basis for the complex of measures designed to protect forests in areas subject to pollution. Determination of the need for an emission reduction is based on a system of special ecological norms that are transformed into the admissible emission requirements with which the enterprises have to comply.

The spatial extent of stand damage around point sources can be related to the annual volume of emissions by industries. The area of any industrial desert that forms is directly related to the concentrations of pollutants and the duration of the emissions. The growth of such deserts depends on the nature of the emissions and any remedial measures that are taken to reduce the impacts of the pollutants. However, once established, industrial deserts are very difficult to rehabilitate, especially in the absence of any emission reductions. This is particularly true in the far north, where the problems are exacerbated by adverse climatic conditions. New industrial point sources can result in the formation of an industrial desert, and the desert will continue to expand until a balance is reached between the strength of emissions and distance from source. As an example, the initial damage to forest stands in the Bratsk region recorded in the late 1970s affected about 100 ha of forest. By 1990, this had spread and the total area of impacted forest stands amounted to 90,000 ha (Martynuk and Kasimov, 1993).

Industrial emissions are a major factor contributing to the considerable changes in the forest biological and geological coenosis within an industrial desert. The area and development dynamics of the desert are governed by emission volume and composition and by the ecotype sensitivity. Consequently, an efficient system of measures aimed at conservation and rehabilitation of damaged forests should be based on the optimal regulation of pollution impacts on the forests during their entire growing cycle.

4.1 Critical levels and loads for pollutants

The critical levels for gaseous pollution are still to be determined for many eastern forests. For general assessments, relative standards are normally used, with the most phytotoxic to least phytotoxic being listed as: $HF > Cl_2 > SO_2 > NO_2 > NH_3 > H_2S$.

Sanitary hygienic norms (Anon., 1984a) have already been developed, and similar ACLs have been developed for forest species in Russia. The ACL rates were first approved and applied in the operational zone of the Schekin Chemical Integrated Enterprise in order to study the impacts of pollution on the Leo Tolstoy House Museum 'Yasnaya Polyana'. In accordance with the forest ACL requirements adopted in Russia, the admissible average daily and peak pollution concentrations in the atmosphere are regulated.

By linking health and forest pollution regulations, standard Russian methods can be used, as can the equipment operated in special hydrometeorological laboratories. A further advantage is that existing unified methods for the calculation of admissible emissions as applied to forests can be adopted. For example, a list of the substances included in the first forest ACL standards developed for the Leo Tolstoy House Museum 'Yasnaya Polyana' consisted of 13 substances (Table 6.3). The majority of ACL standards are lower than those developed for human health. In 1985–1990, the forest ACL values were specified for coniferous species (Table 6.3). In addition, the criteria for the

Table 6.3. Standards for critical levels of gaseous phytotoxic substances in the atmosphere for tree species. (Sources: Anon., 1984b, 1990b.)

	Critical level (mg m^{-3})					
	For humans		For tree species at 'Yasnaya Polyana'		For coniferous species (VNIILM – 1990)	
Substance name	Maximum peak	Daily average	Maximum peak	Daily average	Maximum peak	Daily average
Nitric oxide	0.085	0.04	0.04	0.02	0.05	0.02
Sulphur dioxide	0.5	0.05	0.3	0.015	0.35	0.03
Fluorine gaseous compounds (calculated as fluorine)	0.02	0.005	0.02	0.003	0.006	0.0004
Ammonia	0.2	0.04	0.1	0.04	0.7	0.04
Carbon monoxide	5.0	3.0	5.0	3.0	–	–
Hydrogen sulphide	0.008	–	–	–	0.7	0.1
Carbon disulphide	0.03	0.005	0.008	0.008	–	–
Chlorine	0.1	0.03	0.025	0.015	0.1	0.02
Sulphuric acid aerosol	0.3	0.1	0.1	0.03	15–20	–
Suspended substances	0.5	0.05	0.2	0.05	–	–
Methanol	1.0	0.5	0.2	0.1	–	–
Benzol	1.5	0.1	0.1	0.05	–	–
Formaldehyde	0.035	0.003	0.02	0.003	–	–
Cyclohexane	1.4	1.4	0.2	0.2	–	–

recognition of pollutants affecting ground vegetation and aquatic ecosystems in ecological emergency zones in Russia have been developed and brought into effect (Table 6.4) (Anon., 1992a).

More recently, work has been undertaken in Russia on the development of admissible critical loads for acidic deposition, particularly as related to sulphur and nitrogen compounds. The average load of sulphur in the European part of Russia is 8 kg ha^{-1}, whereas it is 3.5 kg ha^{-1} in the Asian part. Average nitrogen loads are 5.0 and 0.5–3.0 kg ha^{-1}, respectively (Anon., 1996, 1997).

Critical loads for combinations of sulphur and nitrogen calculated in a number of European countries are 5–20 kg sulphur ha^{-1} for coniferous forests and 10–40 kg ha^{-1} for broadleaved forests; the respective values for nitrogen are 3–15 kg ha^{-1} and 5–20 kg ha^{-1} (De Vries *et al.*, 1993). The critical load depends on the site's geology and forest soil type and the geographical location of the forest stand. However, it seems that the critical loads for sulphur and nitrogen deposition are being exceeded in many of Russia's forests. The application of the critical loads theory to Russian forests is still in the research phase (Anon., 1996, 1997; B.N. Moisseyev, personal communication, 1997).

Table 6.4. Critical levels of pollutants in Russia affecting ground vegetation and water systems.

		Parameters			
Parameter number	Indicators	Ecological disaster (Art. 59)[a]	Ecological emergency (Art. 58)[a]	Norm	Time period
Critical levels for ground vegetation ($\mu g\ m^{-3}$):					
1	Sulphur dioxide	> 200	100–200	< 20	Yearly average
2	Nitrogen dioxide	> 300	200–300	< 30	Yearly average
3	Hydrogen fluoride	> 20	10–20	< 2–3	Long-term influence
4	Ozone	> 1500	1000–1500	< 150	Maximal during 1 h
5	Ozone	> 600	400–600	< 60	Average during 3 h
6	Ozone	> 500	300–500	< 50	Average 0900–1600 every day in the period from 1 April to 30 September
Critical loads for forest and water ecological systems ($g\ m^{-2}\ year^{-1}$), northern and central regions:					
7	Sulphur compounds	> 5.0	3.0–5.0	< 0.32	
8	Nitrogen compounds	> 4.0	2.0–4.0	< 0.28	
9	Hydrogen ions	> 300	200–300	< 20	

[a]Articles of Russian Federation Law 'About defence of natural environment' (Chapter VIII, 'Extreme ecological situations').

The effects of acidic deposition may pass unnoticed at first but are evident through increased soil acidity, increased mobility of heavy metals and toxic aluminium and the leaching of the calcium, potassium and magnesium cations needed by plants. In time, photosynthesis rates and forest productivity decrease, and the resistance of the forest stands to unfavourable environmental influences diminishes. The importance of the problem is illustrated by the presence of acidic deposition long distances (> 1000 km) from the sources.

A monitoring network to assess the transboundary spread of pollutants exists in Russia. This network is part of the international Programme of Monitoring and Assessment of Air Pollution in Europe. The network has shown that the average annual values of wet and dry deposition of sulphur in various regions of north-west Russia amounted to 300–680 mg m^{-2}. The values for nitrogen were 120–470 mg m^{-2}. The proportion of dry deposition in the total deposition is increasing. Over the last 8 years, atmospheric pollutant concentrations and acidic deposition away from point sources have both been fairly low and are unlikely to cause large-scale negative ecological effects in the north-west region of Russia. However, when this regional deposition is added to local deposition from point sources, greater problems are likely to occur: an area of 150,000 km^2 may be affected by such a combination.

Soil acidification increases the mobility of heavy metals and stable organic combinations (cf. Binkley *et al.*, 1989). Preliminary data released by the Meteorological Synthesizing Centre 'Vostok' suggest that the influx of transboundary lead and cadmium to Russia's territory from other Transboundary Air Pollution Convention member countries exceeds the exports of these metals from Russia. The incoming amounts of lead and cadmium from Poland, Germany and Sweden exceed the export of iron and lead to those countries by a factor of more than 10. The incoming amount of lead from the Ukraine, Belorussia and Latvia exceeds Russia's exports to those countries by a factor of 5–7. The import of cadmium from Ukraine, Belorussia, Finland and Latvia exceeds exports to those countries by a factor of 7–8.

The lead deposition to the European Territory of Russia (ETR) consists of considerable amounts derived from outside Russia: 1100 t from the Ukraine, 180–190 t each from Poland and Belorussia, and over 130 t from Germany. The deposition of cadmium in the ETR includes 40 t $year^{-1}$ from Ukrainian sources. The deposition of lead and cadmium is of special significance for the western regions of Russia, and 30% of the total heavy metal pollution in the ETR is derived from foreign sources. Consequently, the transboundary transport of heavy metals contributes to the pollution of the Russian territory through the deposition of lead and cadmium (Anon., 1997).

Within the framework of the Convention on Long-range Transboundary Air Pollution, critical loads for acidic combinations of sulphur and nitrogen for forest soils have been calculated for the whole Russian territory, subdivided into a grid of 150 km × 150 km squares. Depending on soil type, stand productivity and soil hydrology, the critical loads for acidity vary from 350 to 4900 g H^+ ha^{-1} $year^{-1}$. Models have been developed by the Institute for Global Climate and Ecology (Anon., 1992b) showing bulk acidic deposition for the whole Russian territory (B.N. Moisseyev, personal communication, 1997).

A second map has been developed that shows the areas at risk from forest damage as a result of the critical loads being exceeded. The degree of risk can be calculated by dividing the potential deposition of acidity value by the critical load, with the resulting quotient indicating the degree of risk. In the Asian part of Russia, the greatest risk of acidification occurs in the forest stands of Ekaterinburg and Chelabinsk *oblasts* (regions). There is a moderate degree of risk in the Vladivostok city area. A relationship has been established between areas with acidified soils and areas subject to high levels of forest pests and diseases (B.N. Moisseyev, personal communication, 1997).

Currently, efforts in Russia to reduce the adverse influences of anthropogenic factors on the environment are focused on the creation of the Unified System of State Ecological Monitoring (USSEM), led by the Federal Forest Service of Russia. As of November 1996, work aimed at the implementation of the first-stage activities related to the creation of the USSEM had started in 48 administrative territorial and national subdivisions of the Russian Federation. Regional information and analysis centres equipped with modern computer technologies, including data-processing technologies, have been created in 20 Russian regions and are currently in operation. Special stations equipped for receiving and processing atmospheric and remote sensing information related to the solution of nature protection problems have been established in Kurgan, Perm, Orenburg and Irkutsk *oblasts* and the Yamalo–Nenetsky national territorial subdivision.

5 Forest fires and atmospheric pollution

There is a relationship between the resistance of forest ecosystems to atmospheric pollution and the safeguarding of forests against fires, forest pests and diseases. Forest fires have a global impact on environment that can be considered to be both natural and anthropogenic. Forest fires can potentially influence the entire biomass of an area, while smoke emissions to the atmosphere can be compared with results of volcanic activity. In Siberia, there are about 30,000 forest fires per year, affecting on average about 5.1 Mha.

In addition to the documented forest fires, unregistered fires occur in the tundra, forest–tundra and in the steppes. Remote sensing analyses of unprotected areas of tundra, forest–tundra and steppes indicate that the areas subjected to fires in these regions are comparable in size to those damaged by forest fires (Valendik, 1996). The total land area of Siberia affected annually by fire therefore amounts to about 10.1 Mha.

Most fires occur in regions characterized by a greater number of rainless days per year and a greater number of drought years per decade, as well as by greater anthropogenic pressures. A high density of forest fires occurs in regions such as the Ekaterinburg *oblast* and the north-western part of the Tumen *oblast*, the eastern part of the West Siberian Plain, including the Tomsk *oblast*, the northern part of the Novosibirsk *oblast* and Enisei River Valley of Krasnoyarsky Krai, the Angara River Valley (Krasnoyarsk and Irkutsk regions) and the south-western part of the Yakut Republic and Chita *oblast*. In the far east, fire-susceptible areas include the southern parts of the Amur *oblast*, Khabarovsky Krai and the whole of Primorsky Krai, and the north-east and south-east parts of the Magadan *oblast*.

Forest fires result in the release into the atmosphere of great quantities of various gaseous and aerosol mixtures; these significantly influence atmospheric dynamics at both local and global levels. During the 2-month-long fire period in Siberia, around 20,106,000 t of biomass are burned and 2,106,000 t of combustion products are released into the atmosphere. The smoke plumes extend over tens of kilometres. Forest fires occurring in zones that are heavily impacted by pollution can transfer pollutants, including heavy metals and radionuclides, from the biomass to the atmosphere, resulting in the spread of the pollutants to previously unaffected areas. One of the best examples of this process was provided by the fire at Chernobyl. As a result of long-range transport of smoke, the influence of forest fires can extend over very large areas. Normally, smoke from forest fires reaches heights of 5–6 km. However, in September 1957, the smoke generated by large-scale fires in Buratia reached a height of 8 km.

The FIRESCAN experiment conducted in the summer of 1993 on a forest plot located on the island of Bor, 600 km north of Krasnoyarsk city, involved the controlled burning of forest in order to measure and record, for the first time in Russia, the changes in contents of gases and aerosols in the smoke column (Kutsenogy *et al.*, 1996). The experiment established that the majority of aerosols released from the burning zone consist of submicron-size particles. It is for this reason that smoke columns reaching a height of several kilometres can be expected to influence the smoke plume over distances of hundreds to thousands of kilometres.

The 1997 forest fire season data illustrate the fire preventive measures taken annually in Russia, as well as the existing problems associated with forest fires. In 1997, the forest fire situation was extremely complicated and threatening in many regions of Russia, especially during May, July and September. A total of 27,356 fires were registered in the State Forest Fund area. The area burned amounted to 657,500 ha. Of the registered fires, 881 were classified as large fires, and these accounted for 82% of the total burned area. The cost to the Russian economy was estimated at 1.263 billion roubles. The volume of standing timber burned or otherwise damaged was 19.8 million m^3, and a further 12.7 million m^3 of cut wood was also lost.

Preparation for the 1997 forest fire season, involving the arrangement of fire preventive actions and enhancement of forest protection measures, was carried out pursuant to the Decree of 27 March 1997, No. 339 'On Urgent Measures for Guarding Forests against Fires and Protection of Forests against Pests and Diseases in 1997'. The territorial bodies of the Federal Forest Service of Russia, together with special aircraft units involved in protecting forests against fires, took measures aimed at the prevention, timely discovery and extinguishing of forest fires. As a result of these activities, the number of forest fires decreased by 6%, the forest area subjected to fires was almost 3 times less, and the average forest fire area was 31.7 ha (compared with 79.1 ha in 1996).

A total of 1770 fire stations of various types, 364 mechanized units, 787 fire-towers and 2286 watch points was prepared for fire-fighting in 1997. In addition, 3570 locations were organized where fire prevention and fire fighting equipment were concentrated, and a further 7034 fuel and lubricants depots were set up, with a total fuel stock amounting to over 3 Mt. The *leshozes* installed 130,000 km of fire-breaks, and forest management enterprises maintained 503,000 km of fire breaks in 1997. The fire prevention information campaign was more active than in previous years, and was also more efficient in utilizing the mass media. Twenty special aircraft units (bases) and two aircraft sections were used to provide aerial surveillance over a total area of 6815 Mha. With the help of this aerial reconnaissance 7900 forest fires were identified and 5700 fires were extinguished using special aircraft. However, the use of aviation in forest fire protection is decreasing, with 28,928 flight hours in 1997 compared with 41,157 h in 1996.

An important prerequisite for the preservation of forests under the present circumstances is continuous forest fire prevention, which today is particularly important given that the state financing of forestry is inadequate. The Federal State Forestry Department is paying special attention to the prevention and extinguishing of forest fires occurring in areas with radioactive contamination, as such fires are an additional hazard for both forestry workers and people living in nearby settlements, as well as for residents of more distant villages, towns and cities. The issue of forest fires in Siberia is actually a problem of the organization of efficient fire prevention and forest protection activities over the vast and sparsely populated territories when only meagre state funds are available for the purpose. The new market economy conditions are forcing Russian foresters to look for low-budget methods of fire prevention and forest protection activities, as well as revising their views on the damage caused by forest fires and the role of fires in the environment. The safety of human populations and settlements is the main criterion determining the forest protection strategy in Russia. This strategy is consistent with the concept of viewing all fires as natural disasters that must be put out in the shortest time possible, no matter where – or in which ecological systems – they occur.

6 Pests and disease

As of January 1997, pests and diseases affected a forest area of 3.0 Mha in Russia, and the area affected on an annual basis varies from 2.0 to 3.5 Mha. The total area requiring immediate sanitary action was 1.3 Mha. Of special concern was the large-scale expansion of the Siberian pine moth (*Dendrolimus pini*) in Krasnoyarsky Krai, in Irkutsk and Tomsk *oblasts*, in Tuva Republic, in Primorsky Krai, and in a number of other regions. In these regions, there are also concentrations of the nun moth (*Lymantria monacha*); the Asian variety of this pest, which is a quarantine species for a number of Pacific Region countries, is also spreading. About 4% of the Russian forest areas is believed to be dead, dying or severely weakened as a result of forest pests. The forest area requiring strict controls as a result of pest and disease activities amounts to less than 11% of the forested area of the State Forest Fund. However, in order to have an accurate picture of the condition of such stands by the year 2000, it would have been necessary to examine around 110 Mha annually between 1997 and 2000.

In 1997, action was taken to eliminate forest pests from an area of 421,164 ha. This effort was largely successful, achieving an efficiency of 80–97%. Both biological and chemical methods were used. Biological methods (application of bacteria and virus preparations) that are safer for the environment than chemical insecticides are being increasingly used, such that they were expected to account for 60% of the effort by 1998. In order to increase the stability of forests, a change in approach is being adopted. Instead of damage-limitation measures in severely impacted areas, practices are being introduced that maintain pest populations at low levels and prevent large-scale outbreaks.

The forest protection service of Russia is being reorganized following the Resolution of the Gosleshoz Collegium of 18 November 1997. A state institution called the Russian Forest Protection Centre is being established; this will provide a unified methodology and informations approach to forest protection problems and will organize forest health monitoring over the whole of the Russian territory. The reorganization will lead to the gradual formation of specialized structural units within the existing state forest management bodies in Russia. In the 21st century, 14 regional centres for forest health monitoring will be established in Central Russia and in the probable areas of origin of the Siberian pine moth.

As a result of the new Forest Code and the Convention on Biological Diversity, the guidelines relating to forest protection in Russia were revised. The Regulation on forest health monitoring contains a new approach to the management of pests and diseases. Forest health monitoring will be conducted over almost all of the Goslesfund area, except for forest reserves. This differs from the current situation, where forest health monitoring is restricted to observations of the most abundant pest populations. Exceptions include the

Moscow and Kaliningrad Forestry Departments that have developed scientifically based observation systems for their forest health monitoring.

Recently, the Federal State Forestry Department has developed and submitted to the Ministry of Justice of the Russian Federation a new set of Forest Health Rules. These aim at increasing control over forest protection activities, conserving the biological stability of forest stands and preventing inappropriate forest use and damage caused by forest management. The measures and activities planned for 1998 were aimed at controlling pest concentrations and preventing the spread of significant pests, better identification of new pest and disease concentrations, and the furthering of forest protection activities.

Evidence for the success of forest protection activities comes from the figures available for severely impacted areas: 610,000 ha in 1996, 577,000 ha in 1997 and 462,800 ha in 1998. The Federal Programme 'Forests of Russia' included in 1998 the field assessment of the forest pest and disease situation in a forest area of 12.2 Mha, greatly improving the understanding of forest health issues in Russia.

7 Data requirements

As the territory of Russia crosses ten time zones, the country's vastness requires the use of remote sensing methods for assessing the ground surface. For instance, the only source of data on forest fires in the unprotected northern forests and tundra is satellite remote sensing. The base operation unit within the all-Russia system of forest fire monitoring should be a regional subsystem servicing territorial special aircraft units. At the headquarters of every large territorial special aircraft unit, there should be a node for the fast reception and analysis of satellite data.

Many trials similar to those of the FIRESCAN experiment need to be conducted in order to link aerial and satellite assessment of the gases and aerosols emissions from forest fires. Satellite data reception stations for the NOAA–AVHRR have been newly installed at Krasnoyarsk and Khabarovsk and will be of great importance for the real-time identification of fires and for fire reconstruction. The NOAA station that has been installed at Krasnoyarsk, which is equipped with a new system for satellite tracking and with Internet access, will increase the possibilities for collaboration between Russian and other scientists. The facility will enable scientists to access the satellite data and collaborate in the assessment of the scale and impact of large fires in diverse forests, focusing special attention on those individual Siberian regions that formerly had no monitoring. This will require coordinated assessments of the satellite data received with the NOAA–AVHRR equipment, and the distribution of existing technologies and scientific research potential.

Preliminary trials continue, intended to check a specific case in the field with the help of high-resolution images received from satellites, as well as the mapping work based on aerial photographs. Plans exist for the field trials of a new fire detector that has recently been installed on the space platform 'Nature' and which is proposed for the space station 'Mir'. A specialized 'Green Wave' programme complex has been designed for the continuous analysis of satellite information to reveal natural vegetation dynamics and the impact on vegetation of industrial activities and fires (Shevirnogov *et al.*, 1996).

8 Conclusion: sustainable forest management in Russia

Recently, the Federal State Forestry Department has developed and enacted (Gosleshoz order No. 21 of 5 February 1998) the Criteria and Indicators for Sustainable Forest Management for the forests of the Russian Federation. This was done to implement within forestry the concept of Russia's Transition to Sustainable Development, as approved by the President of the Russian Federation (Decree No. 440 of 1 April 1996), the State Strategy of Sustainable Development of Russia developed by the Government (Decree No. 559 of 8 May 1996), the resolutions of the UN Conference on Environment and Development (Rio de Janeiro, 1992), the resolutions of XIX Special Session of the UN General Assembly (New York, 1997), and the Forest Code of the Russian Federation. The Criteria and Indicators will stimulate additional analysis of the management capacity of the state forestry bodies in all administrative territorial and national subdivisions of the Russian Federation. Their capability and readiness to manage Russian forests on the principles of sustainability as recommended by the world community will also be tested.

During 1998, the Criteria and Indicators were adapted for use on regional and local levels. Amendments were made to the Instructions for the State Forest Fund Inventory, enabling foresters to take into account these criteria and indicators when making regular forest inventories, as of 1999.

The Criteria and Indicators include the following:

Criterion 2: 'Maintenance of appropriate health status and forest viability', which includes, in particular, the following indicators relating to the impact of atmospheric pollution on forests:

2.1a. Total area of dead and dying forests due to the influence of adverse factors (annually), including industrial emissions.

2.3. Total quantity air pollutants or their quantity per unit of forested area (every 5 years).

2.4. Forest area characterized by serious defoliation, to be assessed through the use of the relevant UN-ECE methods (within the 500 km zone along Russia's western borders).

Thus Russian foresters have a clear picture of the existing forest management problems relating to atmospheric pollution in the forests of Russia and in the whole of Asia.

Russia's 200-year-old state system of forest management has sufficient capacity and flexibility to cope with the changes associated with the new problems emerging in forestry. It can readily take on new tasks and objectives, especially those relating to the preservation of forest viability. The Gosleshoz activities in this important field enable the following aims to be completed in a timely fashion:

- To identify the impact of anthropogenic and natural factors on forests and obtain data about trends and spread of any impacts on Russia's forests.
- To assess the problems related to the maintenance of an appropriate health status and viability of forests, and for the conservation of environmental conditions required for the survival of forest-dependent plant and animal species.
- To determine the course of action needed by Russian national forestry to reduce the negative impacts of atmospheric pollution on the health and viability of Russia's forests.

References

Anon. (1984a) *Critical Admissible Concentrations of Polluting Substances in the Atmosphere of Populated Areas.* USSR Health Ministry, Moscow.

Anon. (1984b) *Temporary Norms for Critical Admissible Concentrations of Polluting Substances in the Atmosphere Causing a Negative Impact on Forest Stands in the 'Yasnaya Polyana' House-Museum Region.* All-Russian Research Institute of Silviculture and Forestry Mechanization, Moscow.

Anon. (1990a) *Report on Contamination and Death of Forests from Industrial Atmospheric Pollution, as of October 1, 1990.* USSR Statistics Department, Moscow.

Anon. (1990b) *Development of Preliminary Norms for Critical Admissible Concentrations of Toxic Substances for the Main Forest-Forming Species.* Final Report on Topic III. 9. 1. for 1990, NIR Leader E.B. Boronin, Moscow, USSR. Goskomles, VNIILM, Moscow

Anon. (1992a) *Criteria for Assessment of Environment Status to Identify Ecological Emergency and Ecological Disaster Zones.* Restricted use circular issued by the General Direction of Science and Technology of the former Ministry of Environmental Protection and Natural Resources of the Russian Federation on 11 November 1992.

Anon. (1992b) *Survey of Background Environment Status in 1991.* Hydrometeoizdat, Moscow.

Anon. (1994) *Survey of Sanitary Status of Russia's Forests in 1993.* Gosagroservice Publishing House, Moscow.

Anon. (1996) *Public Report: On the Environment Status of the Russian Federation in 1995.* International Projects Centre, Moscow.

Anonymous (1997) *Public Report: On the Environment Status of the Russian Federation in 1996.* International Projects Centre, Moscow.

Bezuglaya, E. and Berland, M.E. (eds) (1983) *Climatic Condition Characteristics of Atmospheric Admixtures Spreading. Reference Book.* Hydrometeoizdat, Leningrad.

Binkley, D., Driscoll, C.T., Allen, H.L., Schoeneberger, P. and McAvoy, D. (1989) *Acidic Deposition and Forest Soils. Context and Case Studies in the Southeastern United States.* Springer-Verlag, New York.

De Vries, W., Posch, M., Reinds, G.J. and Kämäri, J. (1993) *Critical Loads and Their Exceedance on Forest Soils in Europe.* Report 58, Winand Staring Centre for Integrated Land, Soil and Water Research, Wageningen, The Netherlands.

Kharuk, V.I. (1999) Air pollution impacts on subarctic forests at Noril'sk, Siberia. In: Innes, J.L. and Oleksyn, J. (eds) *Forest Dynamics in Heavily Polluted Regions.* CAB International, Wallingford, UK, and IUFRO, Vienna, pp. 77–86.

Kharuk, V.I., Nilsson, S. and Samarskaia, E. (1996) *Anthropogenic and Technogenic Stress Factors to Forests in Siberia.* Working Paper 96–104, International Institute for Applied Systems Analysis, Laxenburg, Austria.

Kukuev, Y.A. (1997) The forest resources of the Russian Federation and their regional characteristics. In: Anttonen, T. and Petrov, A.P. (eds) *Potential of the Russian Forests and Forest Industries.* Research Notes 61, Faculty of Forestry, University of Joensuu, Finland, pp. 7–16.

Kutsenogy, P., Valendik, N., Bufetov, N.S. and Baryshev, V.B. (1996) Emissions from large forest fires in Siberia. *Siberian Ecological Journal* 1, 93–101.

Martynuk, A. and Kasimov, V.D. (1993) *Industrial Pollution-Related Damaged Forest Hotbeds and Forestry Strategy under Conditions of Environment Pollution.* VNIILM, Moscow, pp. 3–18.

Nilsson S. (1987) State-of-the-art report on the extent of forest decline attributed to air pollutants in Europe – fall 1986. In: Kairiukstis, L., Nilsson, S. and Straszak, A. (eds) *Forest Decline and Reproduction: Regional and Global Consequences.* IIASA/SRI/IUFRO Workshop, Krakow, Poland, 1987/03/23–28. IIASA, Laxenburg, Austria, pp. 409–418.

Nilsson, S. and Shvidenko, A. (1998) *Is Sustainable Development of the Russian Forest Sector Possible?* IUFRO Occasional Paper No. 11, IUFRO, Vienna.

Rigina, O. and Kozlov, M.V. (1999) The impacts of air pollution on the northern taiga forests of the Kola Peninsula, Russian Federation. In: Innes, J.L. and Oleksyn, J. (eds) *Forest Dynamics in Heavily Polluted Regions.* CAB International, Wallingford, UK, and IUFRO, Vienna, pp. 37–65.

Schlesinger, W.H. (1997) *Biogeochemistry. An Analysis of Global Change.* 2nd edn. Academic Press, San Diego.

Shekhovtsev, A., Zhiltsov, E.V. and Khizhov, S.G. (1997) *Influence of Industries of the Russian Federation on the Environment Status in 1993–1995.* Meteorology and Hydrology Publishing Centre, Moscow.

Shevirnogov, P., Sukhinin, A.I., Kashkin, V.B., Sidko, A.F. and Vysotskaya, G.S. (1996) Scientific Program 'Green Wave' as a means for studying vegetation dynamics in Krasnoyarsky Krai by satellite. *Siberian Ecological Journal* 5, 363–372.

Shvidenko, Z., Venevsky, S.V. and Nilson, S. (1997) Generalized Estimation of Increment and Mortality in Russian Forests. *Proceedings of the VII MAIBL Annual Conference of the IBFRA – 'Sustainable Development of Boreal Forests'*, Moscow: ARICFR, pp. 184–191.

Spiecker, H., Mielikäinen, K., Köhl, M. and Skovsgaard, J.P. (eds) (1996) *Growth Trends in European Forests.* European Forest Institute, Joensuu, and Springer, Berlin.

Strakhov, V.V., Filiptchouk, A.N. and Mokritzki, V.A. (1997) *Forest and Forest Products Country Profile (Condition and Utilization of Forests in 1995) Russian Federation.* Geneva Timber and Forest Study papers No. 14, United Nations, New York.

Valendik, E.N. (1996) Ecological aspects of forest fires in Siberia. *Siberian Ecological Journal* 1, 1–8.

Vassilieva, N.P., Gytarsky, M.L. and Karaban, R.T. (1999) General characteristics of the damage to forest ecosystems in Russia caused by industrial pollution. In: Innes, J.L. and Oleksyn, J. (eds) *Forest Dynamics in Heavily Polluted Regions.* CAB International, Wallingford, UK, and IUFRO, Vienna, pp. 27–36.

7

Forestry Problems and Air Pollution in China and Korea

H. Shang,[1] Y.-K. Kim[2] and D. Xu[1]

[1]Research Institute of Forest Ecology and Environmental Science, Chinese Academy of Forestry, Beijing, China; [2]Air Pollution Laboratory, Forestry Research Institute, Seoul, Korea

Air pollution is becoming an increasingly serious problem in China and Korea. The injuries caused by acidic deposition and dust are greater than those caused by other pollutants in the region. This paper summarizes the information available on the effects of acidic precipitation on the growth and productivity of forest trees and on forest soil, soil microbes, plant diseases and insect pests in China and Korea. Forest decline has occurred in some areas and is closely related to the direct effects of sulphur dioxide, acidic mist or extremely acidic rain events.

In the past decade, acidic precipitation has occurred over extensive areas of China along with increases in the emissions of SO_2 and NO_x. After Europe and North America, China has become the third most polluted region of the world, with extensive areas affected by acidic deposition. The most seriously polluted areas are in central and south-west China. Acid precipitation is also a widely distributed and fairly serious problem in southern and coastal regions. The most acidic precipitation occurs in the extensive regions to the south of the Yangtze River and east of the Sichuan basin, ranging from 31.6 to 100 $\mu eq\ l^{-1}$ (pH 4.0–4.5). In three sub-regions of this area, H^+ concentrations were above 100 $\mu eq\ l^{-1}$ (pH 4.0). In 11 provinces seriously affected by acidic precipitation, the reduction of timber growing stock and the drop in production of the crops attributable to acidic deposition brought about direct economic losses of 4400 and 5100 million yuan (RMB) (*c.* US$5 and 6 million), respectively.

Since the latter half of the 1960s, environmental pollution, industrialization and urbanization have become major social problems in Korea.

 Air Pollution and the Forests of Developing and Rapidly Industrializing Regions (eds J.L. Innes and A.H. Haron)

Current air pollution concentrations in urban and industrial areas are twice that of mountain areas. Today's average precipitation acidity in industrial and urban areas ranges from pH 4.9 to 5.3, which is lower than the pH 5.6 precipitation encountered in mountain areas. About 43% of the total measured precipitation in Korea was below pH 5.6 in 1996. Soil acidity is also affected by air pollution, and soil pH is lower in urban and industrial areas. According to the results of a forest decline survey in 1996, slight damage to forests is present in the southern parts of coastal areas and industrial areas, and in large cities. Large areas of forest decline have not yet been found in Korea.

1 Introduction

Air pollution is an increasingly serious problem in China. Smoke emissions are the most important of the various pollutants; injuries to ecosystems caused by acidic precipitation and dust have been greater than the injury caused by other pollutants in China (China Environmental Status Bulletin, 1996). In the past decade, acidic precipitation has occurred over extensive areas in China, along with increases in the emissions of sulphur dioxide (SO_2) and nitrogen oxides (NO_x). After Europe and North America, China is now the third most polluted region in the world. Large areas are impacted, and the extent of the impacted areas is increasing. The most seriously polluted area is central China, followed by south-west China. Acidic precipitation is also frequent and fairly serious in southern and coastal regions. Air pollution is spreading from cities to rural areas, and the potential for serious ecosystem disruption is increasing.

Korea is classed as a rapidly industrializing country. Cities and industrial areas are growing as a result of population increases and economic growth and the increased industrialization has been accompanied by rising environmental pollution. The SO_2 concentrations in Seoul, for instance, can be compared to those of large European and American cities. Such air pollution also has negative effects on vegetation and forests (Kim, 1996). The amounts of acid-forming pollutants emitted from the expanding cities and industrial areas will increase as the economy develops, so that the damaging effects of acidic precipitation will be severe and will spread to wider areas. On the other hand, the possibility of pollutants derived from distant sources in Far East Asian countries experiencing rapid industrialization needs to be determined and evaluated appropriately in order to cope with their effects on the domestic environment and ecosystems (Kim *et al.*, 1988).

The first environmental pollution accident in Korea occurred during 1930–1940 in Heungnam, when soot and SO_2 from the Heungnam fertilizer factory damaged the forests near the factory (Yun, 1987). Evidence of forest damage induced by air pollution has been detected around industrial areas since late 1960, but the efforts that have been made to reduce damage have been trivial. There is an urgent need to counteract the pollution that is expected to come with expanding industry (Forestry Research Institute, 1994). It is very important to establish effective and modern strategies to combat the influences of air pollution and acidic precipitation, from both the national and international points of view in Far East Asia.

This chapter describes the present status and trends of acidic deposition, its impacts on forests in China and Korea, and the corresponding countermeasures. Nationwide research projects, major research fields, control policy and mitigation measures are also discussed.

2 Forest resources and environment in China and Korea

2.1 Forest resources in China

The area under forest cover in China in 1993 was estimated at 13.9% of the total territory; current FAO statistics give the forest area as 133 Mha (14.3% of the land area) with a further 28 Mha (3% of the land area) of 'other wooded land'. The forests in China are largely distributed in the eastern part where precipitation exceeds 400 mm, including the south-east and south-west monsoon areas. The latitudinal range of the country is substantial, from 18°N in the south of Hainan Island to 53°N in the north-east part of Da Xinganling. The vast area of north-west China has an arid climate and there is only very limited forest cover, primarily restricted to mountainous areas. The climate zones, together with the regions and main tree species, are listed in Table 7.1.

2.2 Forest resources in Korea

The Korean Peninsula is bordered to the north by China and Russia and to the south-east by Japan. The land area of Korea is 221,487 km^2 and is politically partitioned into south and north: the (southern) Republic of Korea has a land area of 99,221 km^2, whereas the (northern) Democratic People's Republic of Korea has 122,041 km^2. Mountains cover most of the north central region, with the Taebaek range extending along the eastern coast. Korea has a temperate climate characterized by summer monsoons and by continental winters with freezing weather. Seasonal changes are gradual but distinctive; spring and autumn are relatively short while summer and winter are rather long. The annual mean temperature ranges from 3 to 16°C and the annual precipitation ranges from 600 to 1600 mm. Granite and gneiss comprise more than 70% of all parent geology. Because of the changeable continental climate and torrential summertime rain showers, the soil is susceptible to rapid erosion. The majority of the forest soil is acidic sandy loam.

Table 7.1. Forest composition in China based on the national forestry inventory (1993).

Region[a]	Forest category	Forest species	Area (10^3 ha)
North-east	Temperate coniferous and mixed forests	*Larix gmelini, Pinus koraiensis* and mixed and broadleaf forests	36,554
South-west	Subtropical mountain coniferous forests	*Abies, Picea, Cupressus,* etc.	25,964
South	Subtropical evergreen and broadleaf forests	*Schima superba, Castanopsis chinensis*	42,043
North and north-west	Warm temperate coniferous and mixed forests	*Pinus tabulaeformis, Picea* and others	20,004
Tibet	Subtropical mountain coniferous forests	*Picea, Abies,* etc.	3,963

[a]The division into five districts is based on the definition in the statistics of the forestry resource by the Chinese Ministry of Forestry (CMOF) (1994).

As a result of its diverse climatic conditions, Korea exhibits various types of forests, divided into subtropical, warm-temperate and cool-temperate forest zones (Kim, 1996). In South Korea (the Republic of Korea), as of 1995, forested land comprised about 6.5 Mha, representing 65% of total land area. A further 7.4 Mha are located in the Democratic People's Republic of Korea. The forested land per capita is very low (0.2 ha per person), being only one-quarter of the world average. The average forest stocking in South Korea is estimated to be 48 m^3 ha^{-1} (Forestry Administration, 1997).

3 Present status of air pollution in China and Korea

3.1 Acidic deposition in China

In China, about 18 Mt of sulphur oxides and 7.5 Mt of nitrogen compounds are emitted to the atmosphere annually, primarily because coal is the dominant source of energy. The majority of this pollution is deposited within the borders of China. Serious effects of acidic precipitation have been observed, particularly in the south-west region, a new industrial area that is exploiting extensive high-sulphur coal deposits (Wang and Shi, 1993). Energy demands and the development of industrial activities are expected to grow more rapidly in the future, and may well result in increased environmental pollution problems.

Brief history of national studies on acidic deposition in China

In 1950, almost 90% of the people in China were living in rural areas, and life was based mainly on agricultural cultivation. Levels of industrial production in the cities were very low, and there were almost no air pollution or acidic precipitation problems. Later, the Chinese government implemented plans for the development of a national economy based on the development of industrial production. In the late 1970s, a major phase in the development of the national economy and industry started and, since then, the rapid economic and social development has been the focus of global attention. During this period, the growth of various industries, transportation and population has led to major increases in the demand for energy. The combustion of fossil fuels, such as coal, oil and natural gas, has increased the emissions of air pollutants (Cheng, 1993; Quan, 1993).

Studies of acidic deposition in China were initiated in the Sichuan province (located in south-west China) and monitoring was occasionally carried out before 1982. After 1982, the National Environmental Protection Agency (NEPA) started a national monitoring project with the objective of obtaining an overview of acidic deposition in China (Zhao and Sun, 1986). A nationwide project on acidic deposition, the 'Source of Acidic Deposition in China and Control Policies', started in 1985. This was supported by the State Planning Committee and was chaired by the NEPA. The project included five sub-projects, detailed below.

The results of the sub-projects showed that acidic deposition was mainly restricted to some areas of south-western and southern China. In some regions, the problem was very serious and substantial economic losses were occurring. Concern about the situation led the State government to invest 5.5 million yuan (RMB) (c. US$700,000) in a continuation of the study, especially focusing on south-western and southern China. This directly led to the project 'A Study on Acid Rain', one of the key projects of the State's seventh 5-year plan (1985–1990). The project was composed of two sub-projects, including the 'Study on acid rain in South-western China' and the 'Study on acid rain in Southern China'. Both sub-projects included five topics (Wang and Shi, 1993):

1. The present status of acid rain in the project area.
2. The transport of acidic pollutants.
3. The chemical process of acidic deposition.
4. The ecological effects of acidic deposition.
5. The control policy for acidic deposition.

By completing these two sub-projects, China obtained information on the monthly and annual distribution, dynamics and trends of acidic deposition on a nationwide scale. In some areas experiencing serious problems from acidic deposition, studies on the sources of acidic pollutants, their transport and their ecological effects were also emphasized. The same survey was carried out in central and eastern China. In 1990–1995, several interdisciplinary and interministry research networks were established.

In addition to the nationwide research projects, many other topics are under study. Three major fields include: (i) distribution, dynamics and trends of acidic deposition; (ii) physical and chemical processes; and (iii) ecological effects and economic losses.

Since 1982, more than 300 monitoring stations and about 900 field points have been established in 26 provinces and autonomous regions all over the country, with the exceptions of Xinjiang, Tibet and Taiwan. The main objective of the acidic precipitation monitoring network is to measure the concentrations of the components in the precipitation.

In 1986, after several years of research, *State Standard Methods for Monitoring and Analysis of Chemical Components in Precipitation* was published. It established procedures for the location of field points, sample collection and transport, monitoring and analytical instruments, and standard methods for laboratory analysis. Quality control and assurance systems were established and regular assessment of the work at monitoring stations was conducted.

Present status of acidic deposition

As indicated above, several nationwide projects have identified the status of acidic deposition in China. The results show that acidic deposition is mainly distributed to the south of the Yangtze River, especially in four regions: the Sichuan basin; Guizhou Province; Hunan, Hubei and Jiangxi provinces; and the coastal provinces of Fujian and Guangdong. Limited acidic deposition was also recorded in northern and eastern China. Local sources were the major reason for acidic deposition; long-range transport plays a relatively minor role. The area affected by acidic deposition accounts for about 6.8% of the total area of China and the area with serious problems accounts for 11.7% of the total area affected by acidic deposition.

Acidic deposition is a more serious problem in the southern areas than in the north. This is partly because soil pH in the northern and southern parts of China differs, with alkaline soils in the north (pH 7–8) and acidic soils in the south (pH 5–6). The chemical components of precipitation are also very different between the southern and northern areas (Table 7.2). Higher concentrations of ammonium (NH_4^+) and alkali metals occur in precipitation in the north than in the south.

Trends in the average total suspended particles (TSP), SO_2 and NO_x concentrations between 1981 and 1992 indicate that the concentration of TSP in cities has decreased greatly in the past decade, as a result of the effective control of smoke dust from coal-burning. For example, the TSP annual average concentration was 700 $\mu g\ m^{-3}$ in 1981 and dropped to 320 $\mu g\ m^{-3}$ in 1992. Southern cities have TSP annual average concentrations lower than the

Table 7.2. The chemical composition of precipitation in northern and southern cities (μeq l^{-1}) (Xu *et al.*, 1995).

Area	Station	Sample number	SO_4^{2-}	NO_3^-	NH_4^+	Ca^{2+}	Mg^{2+}	SO_4^{2-}/NO_3^-	pH
North	Beijing	28	154.5	39.5	162.8	151.6	12.1	3.91	6.90
	Changchun	34	156.5	21.2	61.3	256.5	51.2	7.38	6.71
	Shenyang	19	398.0	50.3	99.0	305.4	395.3	7.96	6.41
	Xi-an	5	358.1	67.3	275.8	1795.4	66.84	5.32	7.15
	Yantai	2	182.5	22.8	39.1	289.1	20.1	8.00	6.79
South	Chongqing	21	326.6	27.9	151.1	127.8	31.5	11.7	4.21
	Guiyang	4	405.2	27.9	174.3	199.6	65.2	14.5	4.23
	Nanning	29	61.6	4.9	27.7	26.6	1.4	12.6	4.82
	Shanghai	36	153.4	12.6	75.8	104.3	27.9	12.2	4.85
	Hefei	42	141.9	31.8	117.3	110.3	13.7	4.46	4.73

national standards, but in northern cities they continue to exceed the standards (Table 7.3). Concentrations of SO_2 and NO_x have not changed in urban areas (Wang and Shi, 1993).

The major cities where TSP concentrations exceed the Grade 3 national standard (annual average concentration 300 μg m^{-3}; daily average concentration 500 μg m^{-3}) are mainly located along the lower stretches of the Yellow River, in the south-east region of China. In this area, large amounts of coal are used for heating in winter, and the area is also very dry due to a shortage of rain. In contrast, TSP concentrations in the region south of the Yangtze River (Chang Jiang) are quite low, primarily as a result of high rainfall (Wang *et al.*, 1993; Xu *et al.*, 1995).

The distribution of SO_2 concentrations in major cities in 1992 indicated that the most seriously polluted cities were located in the north of China, where SO_2 emissions were also the highest. The cities in the south and south-west of China came second, with the exceptions of some such as Chongqing, Guiyang and Nanning, where sulphur contents in coal are high and dispersion is limited because of the effects of nearby mountains on the local climate.

The annual average concentrations of NO_x in all cities were lower than the national standards, in general below 50 μg m^{-3}, because of the small numbers of cars that are currently found in cities. This situation will change, because the numbers of cars are increasing by 10% annually. Consequently, it will not be possible to dismiss NO_x pollution so lightly in the future.

Air quality in rural areas, including villages and towns, is better than in the medium-sized and bigger cities in China. In general, annual average SO_2 and NO_x concentrations are approximately 10 μg m^{-3}, which is close to levels in Europe and North America, but two to three times higher than in Japan and South Korea. TSP concentrations are much higher, approximately

Table 7.3. Weighted average concentration of ions in precipitation in China (μeq l^{-1}) (Wang and Ding, 1997).

Region	Precipitation (mm)	H^+	NH_4^+	Ca^{2+}	Mg^{2+}	Na^+	K^+	SO_4^{2-}	NO_3^-	Cl^-	F^-	Σ^+	Σ^-	Σ^+/Σ^-
Fujian	1379.3	33.15	70.09	53.3	7.3	14.7	5.1	104.46	14.02	19.25	10.96	182.63	148.69	1.23
Jiangxi	1554.6	32.42	51.25	64.5	11.5	22.5	10.4	100.33	19.62	14.87	16.44	192.55	151.26	1.27
Hunan	1274.0	16.61	81.74	62.9	10.1	8.8	7.8	128.30	18.07	16.33	11.67	188.07	174.37	1.08
Zhejiang	1550.1	20.54	68.47	49.3	9.6	24.5	8.4	109.71	17.98	20.31	8.26	180.75	156.25	1.16
Hubei	1108.2	33.99	100.47	64.8	9.5	8.8	7.3	129.46	22.40	16.36	3.24	224.98	171.46	1.31
Anhui	1019.8	4.30	58.54	65.2	8.6	18.6	10.8	106.79	21.37	24.43	1.02	166.00	153.62	1.08
Jiangsu	1212.1	3.29	60.17	116.7	14.0	29.6	8.9	166.30	20.88	27.95	10.67	232.65	225.81	1.03
Shandong	956.8	0.77	54.42	167.6	52.9	41.4	27.9	161.51	22.64	91.38	21.45	343.05	296.97	1.16
Average	1211.9	18.13	67.89	80.4	15.4	21.1	10.8	125.86	19.62	28.86	10.46	213.84	184.80	1.17

150 μg m^{-3}, because of natural dust and the dry climate in northern and north-western China.

Taking the country as a whole, urban air pollution is closely related to local emissions. The emissions of SO_2 are substantial in those provinces located in the eastern parts of the middle of China; emissions of NO_x and TSP are similar to those of SO_2. Air quality in any city or region is mainly determined by such factors as the type of pollutant, emission amounts and characteristics and the environmental geography; the combination of these is unique to each location. In general, air quality in China is worse in big cities than smaller ones, and the cities on the plains have better air quality than those in mountainous areas. Air quality in predominantly rural areas is fair.

A comparison of the spatial distribution of precipitation pH in China in 1982 and 1992 indicated that areas with low pH have increased over the last 10 years. The most acidic area is located in south-east China (Zhao and Sun, 1986; Wang and Shi, 1993).

In 1992, the National Acidic Deposition Monitoring Network, with a total of 271 stations, came into operation, and the following results were obtained (Wang and Ding, 1997).

IONIC CONCENTRATIONS AND THE MAIN CHEMICAL COMPONENTS OF PRECIPITATION

H^+ concentrations ranged from 0.77 μeq l^{-1} (pH 6.11 occurred in Shandong province) to 33.99 μeq l^{-1} (pH 4.47 occurred in Hubei province), and the average value was 18.13 μeq l^{-1} (pH 4.74) (Table 7.3). Over eight provinces, the SO_4^{2-} concentration ranged from 100 to 130 μeq l^{-1}, except in Jiangsu and Shandong provinces where the concentrations were above 160 μeq l^{-1}. The average SO_4^{2-} concentration for the eight provinces was 125.86 μeq l^{-1}, about three times higher than the average concentrations in Europe, South America and Japan (Wang and Ding, 1997). The main ions in the precipitation were SO_4^{2-} (31.6%), NO_3^- (4.9%), NH_4^+ (17.0%) and Ca^{2+} (20.2%). The average ratio of SO_4^{2-} to NO_3^- in precipitation was 6.4, double that in Europe, South America and Japan, indicating that air quality in China is particularly affected by smoke from coal burning.

THE GEOGRAPHICAL DISTRIBUTION OF H^+ CONCENTRATIONS IN PRECIPITATION

The greatest precipitation acidity occurred in the areas to the south of the Yangtze River and in the east of the Sichuan basin (Figs 7.1 and 7.2), with annual average concentrations ranging from 31.6 to 100 μeq l^{-1} (equivalent to pH 4.0–4.5). There were three sub-regions where average annual H^+ concentrations were above 100 μeq l^{-1} (pH 4.0). The acidity of the precipitation there was probably amongst the highest of anywhere in the world. For example, the maximum annual average H^+ concentration in precipitation in Europe and North America is 63.1 μeq l^{-1} (pH 4.2, 1993), and in Japan the equivalent figure was 39.8 μeq l^{-1} (pH 4.4).

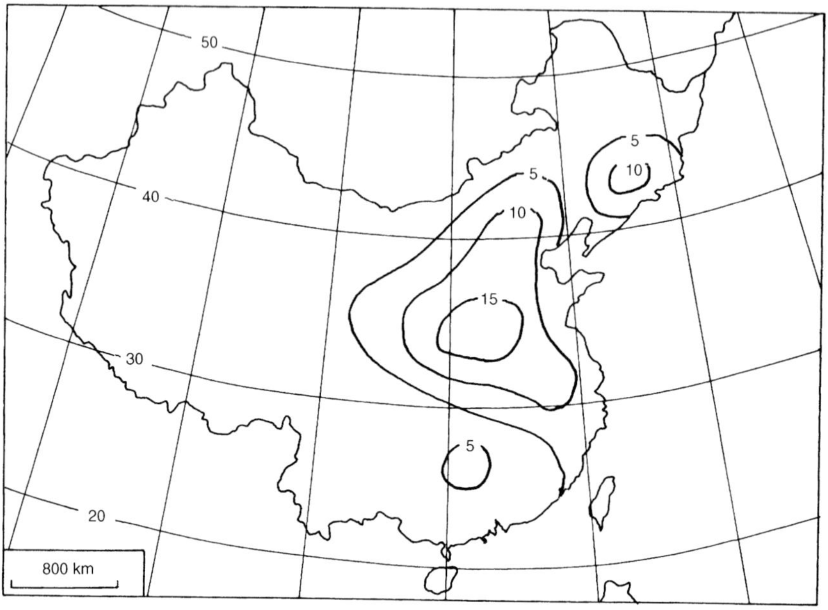

Fig. 7.1. H^+ concentrations (μeq l^{-1}) in precipitation in 1993. (Source: Wang and Ding, 1997.)

THE GEOGRAPHICAL DISTRIBUTION OF NO_3^- CONCENTRATION IN PRECIPITATION

The highest NO_3^- concentrations in precipitation occurred in east-central China, with values of up to 20 μeq l^{-1} (1.26 mg l^{-1}). The geographical distribution of NO_3^- concentrations was similar to that of SO_4^{2-} except for some cities surrounded by high mountains in Guizhou and Sichuan provinces. There, the coal being burnt has a high sulphur content and the local topography limits the dispersion of pollution. Without exception, the areas in southern China characterized by acidic precipitation had low NO_3^- concentrations, with annual average concentrations being lower than the 20 μeq l^{-1} found in areas without acidic precipitation. In general, NO_3^- concentrations in China were lower than in north-east America.

POSSIBLE EXPLANATIONS For a long time, coal has been the major source of energy in China, accounting for 70% or more of the total energy supply; petroleum accounts for about 20% and the cleaner resources, such as natural gas and hydroelectricity, account for only 5% or less (Quan, 1993). This pattern of energy sources is unlikely to change in the immediate future. SO_2, NO_x and the particles released during the combustion of the coal are the major sources of acidified precipitation in China, although the mechanisms for the formation of the acidic deposition are different. Generally, southern areas

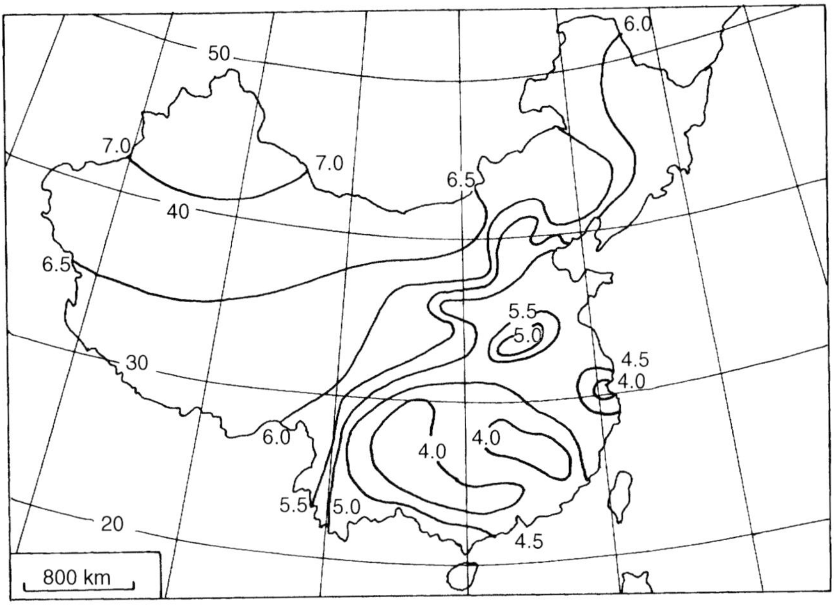

Fig. 7.2. Nitrate (NO_3^-) concentrations (μeq l^{-1}) in precipitation in 1993. (Adapted from Wang and Ding, 1997.)

with serious precipitation acidification and frequent occurrence of acidic precipitation episodes are also the areas with serious air pollution and high atmospheric concentrations of SO_2 (sometimes the proportion of SO_4^{2-} in the anion fraction may be as high as 80%). NO_x is also an important precursor, but its concentration is much smaller in comparison with SO_2. For example, the proportion of SO_4^{2-} to NO_3^- in Guiyang is greater than 10. This is quite different from the situation in European and American countries, where the contributions of SO_4^{2-} and NO_3^- account for 65% and 35%, respectively. The importance of sulphate indicates that acidic precipitation in China is characterized by rain containing sulphuric acid.

Most of the coal-burning boilers in China are not equipped with desulphurization facilities. Washed or desulphurized coal has been used by only a few plants and industries. At the moment, most of the industrial production technologies and facilities in China have a low technological level, with high energy and raw material consumption and low energy utilization efficiency (Table 7.4). About 300 Mt of coal could be saved annually if the energy efficiency in China was as high as that in developed countries such as Japan and the USA. For power plants in China, the coal consumption per kWh is as high as 423 g (national average). If this value could be reduced to 330 g, 80 Mt of coal could be saved and air quality could be improved.

Table 7.4. Efficiency of energy utilization by sectors in developed countries and in China (Xu *et al.*, 1995).

	Efficiency of energy utilization (%)	
	China	Developed countries
Thermal electricity generation	28	35–40
Industrial boilers	55	70–80
Iron and steel production	28	50–60
Nitrogenous fertilizer production	25	50
Railway transportation	8	20–25
Commercial and domestic use	15–18	50–60

3.2 Present status of air pollution in Korea

Energy consumption and the main emissions in Korea

The total consumption of coal and natural gas in 1995 was 51.4 Mt, divided between the different sectors in the following proportions: industry (48%), power generation (33%), residential and commerce (14%) and transportation (5%). The total consumption of petroleum in 1995 was about 648,810,000 barrels, divided between the sectors as: industry (41%), transportation (26%), residential and commerce (22%), power generation (9%) and others (2%) (National Statistical Office, 1995).

Emissions of SO_2 increased from 1.1 Mt in 1988 to 1.5 Mt in 1995. However, there was a significant reduction in the rate of increase of emissions after 1991. This is mainly because high-sulphur anthracite used in the residential and commercial sector is gradually being replaced by town gas. In contrast, NO_x emissions increased rapidly from 416,500 t in 1988 to 924,900 t in 1995. The rate of increase accelerated after 1990 because of a rapid increase in the number of motor vehicles, from 3,394,803 vehicles in 1990 to 8,468,901 vehicles in 1995. The industrial sector is the primary source of SO_2 emissions in Korea, and its importance has been increasing, currently being responsible for 45% of all SO_2 emissions. The second largest source is the power generation sector, and the importance of this is also gradually increasing each year. Emissions of NO_x have increased continuously with time. They are mainly derived from the transportation, industry and power generation sectors (Park, 1996).

Emissions of SO_2 in Korea come from industry (45%), power generation (33%), residential and commerce (11%), transportation (7%) and other sources (4%). NO_x emissions come from transportation (35%), industry (32%), power generation (26%), residential and commerce (6%) and other sources (1%) (National Statistical Office, 1995).

Air pollution monitoring activities in forest areas

SO_2 CONCENTRATIONS Annual mean concentrations of SO_2 were 11.2 μg m^{-3} in the industrial areas, 8.2 μg m^{-3} in urban areas and 6 μg m^{-3} in mountain areas. SO_2 concentrations are below the international standard of 22.5 μg m^{-3} at all sites and do not seem to be at dangerous levels; they have decreased in most urban cities since 1990. This is mainly because of the gradual replacement of high-sulphur anthracite with town gas in the residential and commercial sectors. Nevertheless, an increase of SO_2 concentrations in industrial and mountain areas requires monitoring (Forestry Research Institute, 1993; Kim, 1996).

ACIDIC PRECIPITATION The average acidity of precipitation today in industrial and urban areas ranges from pH 4.9 to 5.3, with levels lower than pH 5.6 in mountain areas. About 43% of the total measured precipitation in southern Korea was below pH 5.6 in 1996. Concentrations of SO_4^{2-} in precipitation were 6.6 mg l^{-1} in urban areas, 4.9 mg l^{-1} in industrial areas and 2.2 mg l^{-1} in mountain areas. Concentrations of NO_3^- in precipitation were 2.8 mg l^{-1} in urban areas, 1.4 mg l^{-1} in industrial areas and 0.9 mg l^{-1} in mountain areas. Sulphate and nitrate are the most important ions contributing to the acidity (Forestry Research Institute, 1997).

CHEMICAL COMPOSITION OF PRECIPITATION UNDER FOREST STANDS Measurements have been undertaken to determine the effects of the canopy on precipitation pH. The pH of open-field precipitation was 4.5, whereas throughfall pH was 4.9 under oriental white oak (*Quercus aliena*) and 4.5 under Japanese red pine (*Pinus densiflora*). The pH of stemflow was 3.8 on oriental white oak and 3.6 on Japanese red pine in urban areas. All trace substance concentrations were considerably higher in throughfall and stemflow than in open-field precipitation. The chemical composition of throughfall and stemflow was dominated by sulphate, with the volume-weighted mean SO_4^{2-} concentrations being 5–36 times higher in throughfall and stemflow than in open-field precipitation. Oriental white oak has more capability to reduce the acidity of precipitation than Japanese red pine (Oh *et al.*, 1987; Kim *et al.*, 1988).

4 Impacts of air pollution on forests

4.1 Impacts of acid rain on the forests in China

Damage to forests may occur because of either direct effects of SO_2 and NO_x or indirect effects involving soil acidification and mobilization of toxic metals such as aluminium. Extremely acidic deposition events may cause direct damage to leaf surfaces.

Direct damage to forest plants

An experiment involving the spraying of simulated acidic precipitation on to 105 woody species has shown that the threshold for acute damage on plant leaves is at a pH value of about 3.0 (Gao, 1987); pH values higher than 3.5 do not normally cause damage. The extent of any damage depends on the number of exposures to acidic precipitation and the duration of those exposures. Leaf injury increases with longer and more frequent exposures. Several field investigations have been conducted in the Ermei Mountains in Sichuan province, where the mean pH value of precipitation and mist has been recorded as being 4.37 and 4.62, respectively. Several symptoms were observed on Forrest fir (*Abies forrestii*), including low needle densities, tip necrosis of needles, premature abscission, branch dieback and reduced radial growth. A similar situation was found for Masson pine (*Pinus massoniana*). About 99% of the needles were green under rain with a pH value more than 5.0, but only 74% in the area where the rain pH value was less than 4.5 (Cao *et al.*, 1989). The chlorophyll *a* content of needles of Masson pine was 45% less than the control when exposed to rain with a pH value lower than 4.5. Exposure to fog with a pH of 4.0 resulted in chlorophyll *b* contents being reduced to 63.3% of the control values (Du and Liu, 1988). The damage symptoms on Masson pine identified in the field and attributed to acidic precipitation are similar to those caused by simulated acidic rain.

Impacts on tree growth

Forest growth rates in China are considered to be much reduced by acidic precipitation. The annual height growth of Masson pine at age 20 exposed to rain with a pH value of 4.3 was reduced by 50% compared with that in the control (Chen and Zh, 1989). Investigations over much larger areas have shown similar trends. A multivariate analysis, based on quantitative theory, of the growth of 20–30-year-old Masson pine forests has been conducted. The variables included in the analysis were rain pH, SO_2 concentration, soil depth, soil pH value, annual precipitation, relative humidity, sunshine hours and accumulated temperature more than 10°C. Of these, the rain pH value was the most important (Feng and Shan, 1986). In another study, Cao and Shu (1990) concluded that rain pH and SO_2 concentration were the most important factors determining annual biomass increment.

4.2 Impacts of air pollution on forests in Korea

Annual biomass production of coniferous stands in Yeochon was 2.1 times higher in non-polluted areas than in polluted areas, but the difference in deciduous stands was only 28%. In the Ulsan area, annual biomass production in different stands depended on sampling sites. The difference in annual biomass production of coniferous stands with a site index of 6 was 60%, but stands with a site index of 10 had no significant difference. In deciduous stands in the Ulsan area with a site index of 16, the biomass production was 27% higher in non-polluted areas than in polluted areas (Kim *et al.*, 1988).

4.3 Impacts of air pollution on forest soils in China

A few simulation studies on the impacts of acidic precipitation on forest soils have been conducted, such as on the copper (Cu) and cadmium (Cd) phases in soils (Xie and Zhou, 1991). Along with the decrease in pH value, Cu and Cd in soil had a significant transformation to the transition phase in yellow-brown and black soils, and to the solvent phase in red soil. Such transformations are considered to have an adverse effect on the growth of plants.

A small catchment about 10 km from the centre of Guiyang has been instrumented for the study of soils, soil water and streamwater chemistry. The molar ratio Al/(Ca+Mg) of soil water is >0.8 in some places. Two small streams have median pH values of about 4.6 and 5.1. Anion adsorption was determined for soils from the Guiyang catchment as well as from a catchment in Nanchang (Liao and Li, 1994). At concentrations corresponding to or somewhat higher than ambient levels in soil water in the Guiyang catchment, the sulphate adsorption was low, typically 2–4 μeq SO_4^{2-} kg^{-1} soil. Some nitrate adsorption was also observed: 2–3 μeq NO_3^- kg^{-1} soil in some cases. Adsorption/desorption may be important in controlling short-term fluctuations of sulphate concentrations in soil water, but is not likely to prevent long-term acidification. Laboratory studies with selected Chinese soils have shown that the anion adsorption is low. The studies confirm that acidic deposition could be affecting soils in parts of south-west China.

4.4 Impacts of air pollution on forest soils in Korea

Soil acidity has also been affected by air pollution in Korea and low soil pH values have been recorded in urban (pH 4.2) and industrial (pH 4.3) areas. In mountain areas, soil pH was significantly lower in coniferous stands than in deciduous stands (Forestry Research Institute, 1997). Forest soil pH increased with distance from the pollution source at a constant rate over short distances (30 km) and at a decreasing rate over longer distances (200 km) (Kim *et al.*, 1988). Soil base saturation was lower in urban and industrial areas. Active aluminium contents in soils increased with soil acidification at the highest rate in industrial areas and also in urban areas.

4.5 Impacts on soil microorganism and pest occurrence in China

Many studies have shown that acidic precipitation can reduce the reproduction of microorganisms significantly. Investigations on microorganisms in forest soils exposed to acidic deposition have shown that their biomass was significantly reduced (Zhou, 1988). The biomass of bacteria was reduced the most, followed by that of N-fixing fungi. Forest soils were exposed to simulated acidic precipitation with pH of 5.6, 4.7, 4.2, 3.7 and 3.1 for 3 weeks. The reproduction of *Azotobacter*, *Actinobacillus* and other bacteria were inhibited at pH 4.7 and less, with the effects becoming more pronounced as the pH dropped. Urease activity (see Dick, 1997, for an account of this and other soil enzymes) showed an inverse relationship with pH treatment, being 80.28, 84.88 and 88.17 mg NH_4^+-N g^{-1} soil sample at treatments of pH 5.6, 4.7 and 4.3, respectively. However, there was a sharp decline in urease activity below pH 4.3. The responses of protease and phosphorite to simulated acid rain varied in soil samples collected from different regions. This indicates that the impact of acidic deposition on soil enzymatic activity is related to factors that include the physico-chemical soil characteristics (Liu, 1990).

An investigation of the diseases affecting Masson pines has found that the incidence of disease is significantly higher in areas where the precipitation pH is below 4.5 than in areas above this value. The incidence of disease in areas severely impacted by acidic deposition was 2.5 times as high as in unaffected areas. However, needle-cast of Masson pine was controlled in areas with a precipitation pH lower than 4.0. A relationship between the percentage of Masson pines infected with the saprogen, the acidity of the precipitation and tree vitality was also identified. The stronger the precipitation acidity and the weaker the tree, the higher the infection percentage was. When Masson pines infected with red rot and brown spot needle blight were exposed to simulated acidic rain at pH 2.0, 2.5, 3.0, 3.5, 4.0, 4.5 and 4.6, it was evident that neither pathogen could reproduce under the more acidic treatments, and that both were killed quickly at pH 2.0. There was severe growth inhibition at pH 2.5, and survival was limited to about 24 h. At pH 3.0, growth inhibitions were still encountered, although survival was better than at lower pH. Rapid growth was encountered at pH 3.5 to 4.0. Consequently, ambient levels of acidic deposition may be within the range at which a number of pathogens are promoted (Yuan, 1988).

Declining Masson pine forests in southern mountains of Chongqing have suffered damage from pine-moth attacks 15 times in the past 30 years. It is believed that there is a relationship between this very high frequency of attack and the presence of air pollution (Yu and Ma, 1989). An investigation conducted in Maochaoba, in Fengjie County of Sichuan province, found that the mortality of Armand pine (*Pinus armandis*) was related to elevation and the high concentration of sulphur oxide gases in the air, due to traditional sulphur smelting methods used in the area. High elevations and high densities of SO_2 were associated with high mortality and a high incidence of pests (Ma and Yu, 1989).

The higher incidence of forest pests in areas impacted by acidic deposition suggests that the trees have been weakened and are less resistant to pests. Pests aggravate the decline of forests, leading to the increased occurrence of other pests in the forest ecosystems (Yu *et al.*, 1985).

5 The evaluation of economic losses caused by acidic deposition

The forest area damaged by acidic precipitation in the Sichuan basin was 275,600 ha, accounting for 31.9% of the forest area. Up to 15,000 ha (5.7% of the forest area) of forest has died as a result of the acidic precipitation. In Guizhou province, 140,500 ha have been damaged.

Acidic deposition has caused growth reductions in the forests of the Sichuan basin, reducing timber outputs by 346,300 m^3 annually. This represents an annual economic loss of 141 million yuan (*c.* US$17.3 million). In Guizhou province, timber production has been reduced by 132,100 m^3 annually, representing an annual loss of 54 million yuan (*c.* US$6.6 million). Preliminary estimates suggest that the forests of the Sichuan basin and central Guizhou province have been affected by acidic deposition since the 1970s. This means that there may have been a reduction in the growing stock amounting to 11.7 million m^3, equivalent to a loss of nearly 5400 million yuan (RMB) (*c.* US$663 million) up to 1997 (Feng, 1993).

Estimates suggest that in the 11 provinces seriously affected by acidic precipitation (Jiangsu, Zhejiang, Anhui, Fujian, Jiangxi, Hubei, Hunan, Guangdong, Guangxi, Sichuan and Guizhou) the reduction in timber growing stock and the drop in crop production caused by acidic deposition could be responsible for direct economic losses of up to 4400 and 5100 million yuan, respectively. Based on a ratio of 1:8 for economic benefits to forest ecological benefits, the losses of ecological benefits could be about 45,900 million yuan every year (Chen and Yin, 1996).

6 Control policies and countermeasures against acidic deposition in China

China is a developing country with a population of more than 1.2 billion, and is faced with serious challenges from environmental pollution associated with the development of its economy. The Chinese government considers environmental protection and pollution control to be basic national policies, and strives for the solution of environmental protection problems while fostering the development of the economy and society. The general strategies adopted by the government are: (i) saving energy, reducing pollutant emissions; (ii) revising the fuel structure of domestic usage; (iii) improving combustion techniques; (iv) strengthening regulation of the discharge of gaseous pollutants from automobiles; (v) strengthening environmental legislation; and (vi) supporting energy generated with low pollution or without pollution.

The Chinese government developed and issued *China: Ten Major Countermeasures for Environment and Development* in 1994, shortly after the UN Conference on Environment and Development (the 'Rio Summit'). In addition, *China Agenda 21* and *Action Plan for China Agenda 21* have been elaborated. These reflect China's strategy of resolving its environmental problems through its own efforts. Population control investments increased to 4.85 billion yuan from 1981 to 1985, accounting for 0.57% of the GNP, and further increased to 10.906 billion yuan from 1986 to 1990, making up 0.63% of the GNP. The annual investment increased fourfold from 1981 to 1990 (Table 7.5) and a further increase will take place.

Annual investment in air pollution treatment doubled between 1985 and 1990 (Table 7.6). As a result, the smoke emissions from coal burning were limited to 14 Mt and industrial dust emissions were cut to 40%.

In February 1998, the Chinese Environmental Protection Agency indicated that, by 2000, industrial pollution sources discharging SO_2 would meet emission standards and would regulate total SO_2 emissions. In certain cities, including the municipalities, the provincial capitals, the special economic zones, the coastal open cities and the key tourist cities, SO_2 concentrations would reach the national standard for environmental quality, and the

Table 7.5. Environmental protection investments in China in relation to GNP (Xu *et al.*, 1995).

Year	Environmental protection investment (billion yuan)	GNP (billion yuan)	Environmental protection investment rate in GNP (%)
1981	2.50	477.3	0.52
1982	2.866	519.3	0.55
1983	3.072	580.9	0.53
1984	3.336	696.2	0.48
1985	4.850	855.8	0.57
1986	7.389	969.6	0.76
1987	9.189	1130.1	0.81
1988	9.998	1398.4	0.71
1989	10.249	1578.9	0.65
1990	10.906	1740.0	0.63

Table 7.6. Waste gas treatment in China (1985–1990). Source: Annual Report on Environmental Statistics (Xu *et al.*, 1995).

Item	Unit	1985	1986	1987	1988	1989	1990
Waste gas treatment funds	Billion yuan	0.73	0.96	1.24	1.53	1.58	1.48
Total emissions	Billion m^3	7397.0	6967.9	7727.5	8238.0	306.5	8542.2
SO_2 emissions	Million tonnes	13.25	12.50	14.12	15.23	15.56	14.95
Smoke dust emissions	Million tonnes	12.95	13.84	14.45	14.86	13.98	13.24
Industrial dust emissions	Million tonnes	13.05	10.75	10.04	11.25	8.40	7.81
New added treatment capacity	Billion m^3	617.70	370.40	19.60	34.30	473.40	42.10

tendency for increasing acidity in precipitation would be halted. By 2010, the total amount of SO_2 emissions should be the same as in 2000. SO_2 concentrations will meet the national standard for environmental quality, and the area within the acid rain control zone with precipitation pH under 4.5 should be significantly smaller than in 2000.

Several new control countermeasures will be adopted. New coal mines where the sulphur content of coal steam is higher than 3% will be prohibited. Existing ones will be gradually shut down, or closed immediately. In cases where the sulphur content of coal steam exceeds 1.5%, new or renovated facilities should incorporate appropriate facilities for washing coal. Existing coal mines without washing facilities should gradually rectify the deficiency. The sulphur content of coal must be within the regulations issued by local governments.

New coal-fired power stations will be forbidden within or in the immediate vicinities of large and medium-sized cities. Desulphurization facilities must be built in the new or renovated coal-fired power stations using coal with a sulphur content above 1%. Measures for reducing SO_2 emissions should have been taken before the year 2000. Enterprises releasing serious pollution, such as the chemical industry, metallurgical industry, building materials industry and non-ferrous metal industry, must build facilities for the disposal of exhaust gases or other measures for reducing emissions. In combination with adjustments to industrial and product structures, the State council has asked for great efforts to be made in implementing policies of production without pollution, for enhanced technical transformation, for promoting resource conservation and for comprehensive utilization, all of which should decrease the levels of SO_2 emissions.

References

Cao, H.F. and Shu, J.M. (1990) Research on economic losses of crop and forest affected by acid deposition in Guangdong and Guangxi provinces. In: Chinese Research Academy of Environmental Science (ed.) *The Colloquia for Celebrating 10th Anniversary of Establishing Chinese Academy of Environment Science*, Beijing, pp. 101–114.

Cao, H.F., Gao, Y.X., Shu, J.M., Liu, Y.Y., Shu, Y., Liu, L.G., Chen, Y.Z. and Wang, W. (1989) The preliminary research on the relationship between *Abies* decline and acidic deposition. In: Chinese Society of Environmental Science (ed.) *Collected Works on Acid Rain*. China Environmental Science Press, Beijing, pp. 409–420.

Chen, C.Y. and Zh, J.W. (1989) The effects of acid rain on productivity of masson pine in Chongqing City. In: Chinese Society of Forestry (ed.) *Acid Rain and Agriculture*. China Forestry Press, Beijing, pp.151–156.

Chen, N. and Yin, Q.H. (1996) The estimates of economic loss caused by acidic deposition in Chongqing city. *Journal of Environmental Science (China)* 17(5), 24–27.

Cheng, Z.F. (1993) State and trend of acid precipitation monitoring in China. In: *Proceedings of the Expert Meeting on Acid Precipitation Monitoring Network in East Asia, Toyama, Japan*. JSCE Toyama, Japan, pp.187–195.

China Environmental Status Bulletin (1996) *China Environment Yearbook*. China Environmental Science Press, Beijing.

Dick, R.P. (1997) Soil enzyme activities as integrative indicators of soil health. In: Panhurst, C.E., Doube, B.M. and Gupta, V.V.S.R. (eds) *Biological Indicators of Soil Health*. CAB International, Wallingford, UK, pp. 121–156.

Du, X.M. and Liu, H.T. (1988) The characteristics of fog and its effects on masson pine in Nanshan Mountain of Chongqing. *Journal of Environmental Science (China)* 8(4), 467–469.

Feng, Z.W. (1993) *The Effects of Acid Rain on Ecosystems. Research on Acid Rain in Southwestern China*. China Science Technology Publishing House, Beijing.

Feng, Z.W. and Shan, Y.F. (1986) The effects of acid rain on Masson pine in Chongqing region. *Atmospheric Environment and Acid Rain* 1(2), 38.

Forestry Administration (1997) *Green Korea*. Forestry Administration, Seoul.

Forestry Research Institute (1993) Movement and management survey method of forest damage by air pollution. In: *Forestry Research Institute Papers 1993*. Forestry Research Institute, Seoul, pp. 3–10.

Forestry Research Institute (1994) *Impacts of Environmental Pollution on Forest Ecosystems*. Ministry of Science and Technology, Seoul.

Forestry Research Institute (1997) Monitoring of forest damage by air pollution. In: *1997 Research Report of the Forestry Institute of Korea*. Forestry Research Institute, Seoul, pp. 215–248.

Gao, X.P. (1987) The responses of 105 species of plant to simulated acid rain in China. *Environmental Science* 7(2), 16–19.

Kim, Y.K. (1996) Luftschadstoffsituation in Waldgebieten in SüdKorea. *Centralblatt für das gesamte Forstwesen* 113(1), 39–49.

Kim, Y.K., Oh, J.H. and Chai, J.S. (1988) *Effects of Air Pollution and Acid Precipitation on the Forest Ecosystem*. Ministry of Science and Technology, Seoul.

Liao, B.H. and Li, C.S. (1994) The effects of acid rain on ecological environment and parent material. *Agricultural and Environmental Protection* 13(4), 179–181.

Liu, L.G. (1990) The effects of acid rain on activity of soil microorganisms. *Environmental Science Research (China)* 4(3), 37.

Ma, G.J. and Yu, S.W. (1989) The preliminary observation on air pollution, acidic deposition, acid fog, and decline of Armand pine in the mountain area of Fengjie County. In: Chinese Society of Forestry (ed.) *Acid Rain and Agriculture.* China Forestry Press, Beijing, pp. 157–162.

National Statistical Office (1995) *Korea Statistical Yearbook.* National Statistical Office, Seoul.

Oh, J.H., Kim, Y.K. and Chai, J.S. (1987) Studies on the changes of pH chemical composition of throughfall and stemflow under coniferous, deciduous forest stands due to acid precipitation. *Forestry Research Institute, Journal of Forest Science of Korea* 34, 159–165.

Park, S.U. (1996) Interannual trend of anthropogenic emissions of SO_2, NO_x, and NH_3 in South Korea. In: *International Seminar on Air Quality Control and Management in Asia Pacific Region.* National Institute of Environmental Research, Seoul, pp. 105–132.

Quan, H. (1993) Introduction to the acid precipitation issue and its control measures in China. In: *Proceedings of the Expert Meeting on Acid Precipitation Monitoring Network in East Asia, Toyama, Japan.* Japanese Society of Air Pollution, Toyama, Japan, pp. 95–105.

Wang, W. and Shi, Q. (1993) Study on air pollution and its control in China. In: *Proceedings of the 34th Annual Meeting of the Japanese Society of Air Pollution, Chiba.* Japanese Society of Air Pollution, Chiba, Japan, pp. 7–12.

Wang, W.X. and Ding, G.A. (1997) The spatial and temporal distribution of precipitation acidity and ions concentration of precipitation in China. *Research in Environmental Sciences* 10(2), 1–6.

Xie, Q. and Zhou, D.Z. (1991) The behavior of Cd and Cu in soil and its toxic effect under simulated acid rain. *Environmental Science* 12(2), 24–28.

Xu, K.Q., Wang, W. and Sudo, R. (1995) Acidic deposition and its control countermeasures in China. *Journal of Global Environmental Engineering* 1, 27–42.

Yu, S.W. and Ma, G.J. (1989) Discussion on reason of decline of Masson pine forest in south mountains of Chongqing. In: Chinese Society of Forestry (ed.) *Acid Rain and Agriculture.* China Forestry Press, Beijing, pp. 139–140.

Yu, S.W., Ma, G.J. and Bian, Y.M. (1985) Save forest – the investigation report for death reason of a large stretch of pine. *Environmental Science* 6(5), 63–65.

Yuan, C.W. (1988) The preliminary report concerning acid rain and disease occurrence of Masson pine. *Ecological Magazine* 7(5), 50–53.

Yun, I.G. (1987) The relationship between environmental and industrial health. In: Chinese Society of Environmental Science (ed.) *Seminar on Environmental Factors Affecting Health and Welfare.* China Environmental Science Press, Beijing, pp. 23–30.

Zhao, D. and Sun, B. (1986) Air pollution and acid rain in China. *Ambio* 15, 2–5.

Zhou, C.L. (1988) The effects of acid rain on activity of soil microbes. *Ecological Magazine* 7(2), 21–24.

Forestry Problems in South-east Asia 8

T. Dahlan and D.M. Puat

Regional Centre for Forest Management (RCFM), Kuala Lumpur, Malaysia

The forest fire issue has been widely discussed in the international arena. Episodes of haze associated with forest fires over the past few years have captured global attention, as they pose a significant health hazard and cause disruptions to ecosystems. This chapter describes the status, causes and extent of forest fires in the various countries in the South-east Asian region. The impact of haze on tropical forests is also well-known. The policies and legislation of member countries pertaining to forest fire and haze as well as the future challenges that they face are also highlighted. These include changes in land and forest management practices, education, law enforcement and transparency. The issues of forest fire and haze as well as the future challenges culminate in the need to develop further regional cooperation towards a coordinated approach to combat forest fires.

1 Introduction

The South-east Asian region, in its widest sense, embraces not only the Association of South-East Asian Nations (ASEAN), but also Cambodia, Laos and Papua New Guinea. The region is the most populous and developed part of the tropics and, moreover, experienced by far the most rapid economic growth between the 1970s and mid-1990s. The region was forecasted to have a population in excess of 1.49 billion by 2000 (Aksorn, 1991).

The region is endowed with a vast area covered with tropical forest. Out of the total land area of the region, about 4.5 million km^2, at least 48% consists of tropical forest. However, Collin *et al.* (1991) (cited in Ishemat and Lemmens, 1994) estimated that the rate of deforestation in South-east Asia (Indonesia, Malaysia, Papua New Guinea, the Philippines, Thailand and Vietnam) was approximately 1.7 Mha year^{-1} in the period 1985–1990. The available evidence suggests that the rate of deforestation is substantial, despite the shrinking resource (Table 8.1).

Forests are vital for many reasons, including the conservation of biological diversity, soil stabilization, protection of watersheds, the maintenance of hydrological cycles and climate itself, and also for the continued existence of indigenous cultures. All of these forest functions are already threatened by human activities, such as conversion to agriculture, urban expansion, logging for timber and air pollution. Climate change represents an additional and potentially disastrous new stress. Climate change is likely to push many forest areas closer to the brink of disaster. Much speculation about climate change has been indulged in, but it is becoming clear that tropical forest ecosystems are in great danger and severely threatened.

Lately, many countries have shown an interest in the state of their forests in the form of initiatives at various levels to discuss environmental problems and attempts to resolve these problems or at least to limit their effects. The

Table 8.1. FAO estimates of forest cover and deforestation of tropical rainforest (from Holdgate, 1994).

Geographic sub-region/ region	Number of countries	Land area (Mha)	Forest cover (Mha)		Annual deforestation	
			1980	1990	Million ha	%
Africa	40	2236.1	568.6	527.6	4.1	0.7
West Sahelian Africa	6	528.0	43.7	40.8	0.3	0.7
East Sahelian Africa	9	489.7	71.4	65.5	0.6	0.9
West Africa	8	203.8	61.5	55.6	0.6	1.0
Central Africa	6	398.3	215.5	204.1	1.1	0.5
Tropical southern Africa	10	558.1	159.3	145.9	1.3	0.9
Insular Africa	1	58.2	17.1	15.8	0.1	0.8
Asia and Pacific	17	892.1	349.6	310.6	3.9	1.2
South Asia	6	412.2	69.4	63.9	0.6	0.8
Continental S.E. Asia	5	190.2	88.4	75.2	1.3	1.6
Insular S.E. Asia	5	244.4	154.7	135.4	1.9	1.3
Pacific	1	45.3	37.1	36.0	0.1	0.3
Latin America and Caribbean	33	1650.1	992.2	918.1	7.4	0.8
C. America and Mexico	7	239.6	79.2	68.1	1.1	1.5
Caribbean	19	69.0	48.3	47.1	0.1	0.3
Tropical S. America	7	1341.6	864.6	802.9	6.2	0.7
Total	90	4778.3	1910.4	1756.3	15.4	0.8

effects of environmental problems vary greatly, depending on the causes and the areas where the problems occur. Air pollution, for instance, could produce detrimental effects on a local or even regional scale, the latter as a result of transboundary pollution. The effects may take many forms, including environmental, social and economic impacts.

The haze episodes that occurred in 1982, 1991, 1994, 1997, 1998 and 1999 were associated with forest fires in Indonesia and captured global attention. The fires at the end of 1997 shrouded many parts of the region in haze, including Brunei Darussalam, Malaysia, Indonesia and Singapore. The event was considered as the worst ever to affect the region, as it persisted for several weeks. The haze, resulting from suspended particulate matter released by the fires, was not only a health hazard but also caused disruption to the aviation and marine industries and to agricultural output.

2 Tropical forest cover

To illustrate the extent of the deforestation problem, FAO (1992) produced a set of statistics on tropical forest cover in the world. In 1990, 36% of the tropical land area was forested. The largest extents of forest were in Latin America and the Caribbean (8.4 million km^2: 50% of the land area), followed by Africa (6.0 million km^2: 27% of the land area) and Asia (2.75 million km^2: 31% of the land area). About 75% of tropical forests are confined to two zones: the tropical rainforest zone and the moist deciduous forest zone (Table 8.2). Deforestation was estimated by FAO at 15.4 Mha (0.8%) annually, with the highest rate being in the moist and montane zones (Holdgate, 1994).

Table 8.2. The forest status in the South-east Asian region.

Country	Land area (km^2)	Forest area (km^2)	Percentage of forest area
Brunei Darussalam	5,765[a]	4,691[b]	81.37
Cambodia	181,035[a]	132,271[c]	73.06
Indonesia	1,919,445[a]	1,138,000[d]	59.29
Laos	236,800[a]	111,816[e]	47.22
Malaysia	330,355[a]	188,800[f]	57.15
Myanmar	676,550[a]	293,269[g]	43.35
Philippines	299,765[a]	56,860[h]	18.97
Singapore	636[a]	Nil	Nil
Thailand	513,517[a]	156,600[i]	30.50
Vietnam	331,690[a]	93,022[j]	28.04
Total	4,495,558	2,175,329	48.39

Sources: [a]IUCN (1991); [b]Saidin (1993); [c]Beang (1996); [d]Mursidin (1996); [e]Phengdouang (1995); [f]Kamis and Basri (1996); [g]U Than (1996); [h]Javier (1996); [i]Chudchawan (1996); [j]Nguyen and Le (1996).

3 Forestry issues and problems

During the 1990s, South-east Asia experienced the fastest economic growth in the world. This had produced enormous problems for environmental sustainability. The growth and location of the region's population are key determinants of the haze episodes in the South-east Asia region. As the population increases, the human impact on tropical forest lands is perpetually increasing, causing overall degradation, and conversion of rainforest vegetation to pyrophytic life forms (Goldammer and Seibert, 1990).

The forest area has been reduced dramatically, with the rate of loss exceeding the rate of recovery. This has brought about the danger of biological mass extinctions. Many of the region's forests have been greatly reduced by large-scale land settlement schemes. The expansion of agricultural land use will certainly persist, but its limits are fast approaching. In addition, the reduction in the lengths of the fallow periods will, in the absence of other changes, soon lead to soil degradation in those areas still practising land rotation systems of agriculture. The forest ecosystem everywhere in South-east Asia has already become highly degraded, primarily as a result of human activities of various intensities.

4 Forest status in the South-east Asia region

Aggregated over the whole region, the available information indicates that the forest area is shrinking, from 61% in 1973 to 55% in 1988, and it may fall below 50% during, or soon after, 2000.

Within the region, there are major differences in these trends and, on a regional basis within countries, there are also marked differences in the patterns of deforestation. In Cambodia, Indonesia, Malaysia and Laos, more than 50% of the land was recorded as being forested in 1988. The total forest area appears to be relatively stable in Myanmar and Cambodia. There have been substantial reductions in forest area over the last 15 years in Malaysia, the Philippines, Thailand, Laos and Vietnam. In the Philippines, about 70% of the country was forested at the end of the Spanish period and 57% in the 1960s; by 1988 the forest area had been reduced to 21.5% of the land area (Bautista, 1990). The Philippines used to have 16.5 Mha of forest; now only 5 Mha remain, and of this only 1.2 Mha are virgin forest (Cameron, 1996). In Vietnam, the forest area was reduced from 48% in 1943 to 24% in 1983 (Cuc, 1995).

5 Policies and legislation

The environmental condition of the developing countries of South-east Asia has worsened since the 1960s and is expected to continue to worsen.

Deforestation is perhaps the most serious environmental concern. Between 1980 and 2000, 11% of the forest cover in South-east Asia alone is expected to be lost. At the same time, air pollution is expected to increase, by about five- to tenfold, because of anticipated increases in motor traffic, forest burning for agriculture and industrial activities (Celso, 1986). This grim portrait is a statement of the necessity for strong and effective environmental policies throughout the region.

The successful implementation of a forest policy in each country to protect the region against forest problems is highly dependent upon broad-based support from all sectors of society, particularly civic organizations and groups working with the responsible government authorities. Whatever policies are adopted for forest/environmental protection, it is imperative that they provide for effective and continued monitoring, consistent enforcement, institutionalization within government and administrative flexibility. All too often, and particularly within developing nations, the capacity to enact policy outstrips the ability to monitor and enforce the policy, as both monitoring and enforcement are another form of direct and indirect environmental costs.

However, it is important to note that the governments of at least some South-east Asian countries are aware of global climatic change and its likely implications for human activities. They are making efforts to improve understanding of the nature and mechanisms of regional climate by having sound and clear policies on the environment.

5.1 Brunei Darussalam

The 1978 Wildlife Protection Act provides for the establishment of wildlife sanctuaries. A Wildlife Conservation Enactment has been drafted as part of a Special Report on Wildlife Conservation and Management prepared for the Negara Brunei Darussalam Master Plan 2000 to provide, *inter alia*, for the establishment of national parks, wildlife sanctuaries and nature reserves (Farmer *et al*., 1986). The 1934 Forest Act, together with the Forest Rules of 1955 and 1975, provides for the establishment of forest reserves and allocates the forest into one of four categories: protection forest, conservation area, recreational area and production forest (IUCN, 1991).

5.2 Indonesia

Government policy regarding conservation is based upon the desire to promote the cultural and economic development of the Indonesian people in harmony with their natural environment, as expressed in the 1945 Constitution. According to this policy, all forms of natural life and examples of all ecosystems have an important role in human welfare and must be preserved for the benefit of present and future generations. This is elaborated in the Basic Environmental Law of 1982, which makes provision for the management of the living environment. It emphasizes the importance of forest lands, stating that they form the primary means of maintaining harmony between man and the environment (IUCN, 1991).

5.3 Laos

Laos is one of the few countries in Asia that has neither protected areas legislation nor protected areas. A number of central government decrees dating from 1979 deal with various aspects of nature conservation, including forest protection, wildlife trade and hunting and fishing, but these are not comprehensive and have not been effectively implemented. However, a draft Nature Conservation Act has been formulated and is now under consideration by the government. Apart from that, a comprehensive forestry policy is being formulated by the government and this will include environmental conservation components (IUCN, 1991).

5.4 Malaysia

The Malaysian constitution delegates rights over land matters to the respective state governments. Thus, jurisdiction over most land matters is controlled by state decision-making bodies and land ownership is retained by the states. Sabah and Sarawak each have state laws covering forestry, protected areas and wildlife. Statutes relating to biological resources are in force at both state and federal levels, providing a complexity in approach. However, of the many legislative acts in place, only the Environmental Quality Act (1974) and the Fisheries Act (1985) are applied nationally.

There is no specific legislation on forest fires, except a provision in the National Forestry Act, revised in 1993 (Section 81.1.b), prohibiting fire-related activities in permanent forest reserves and detailing the penalties involved in such offences. Likewise, the Environmental Quality Act of 1974 explicitly prohibits open burning without a permit, with the objective of preventing and reducing air pollution and the haze phenomenon.

5.5 Myanmar

Forest policy recognizes the basic tenets of conservation and has three salient principles: the maintenance of environmental stability for the preservation of natural heritage by conserving species and ecosystem diversity; the establishment of a system of protected areas; and the sustainable utilization of forest resources for the direct benefit of present and future generations.

Legal protection of natural resources currently rests on two Acts, both dating from the pre-World War II colonial period. The 1902 Burma Forest Act repealed all earlier forest acts. This Act allows the Ministry of Agriculture and Forests to establish game sanctuaries and reserved forests on any land at the disposal of the government, and designates responsibility for their management and protection (IUCN, 1991).

5.6 Thailand

The earliest official conservation measures were taken during the 13th century by King Ram Khamkaeng the Great, who established Royal Dong Tan Park. The 1964 National Forest Reserves Act was passed to strengthen the inadequate forest protection afforded by the earlier 1941 Forest Act. This Act defines two forest types: production and conservation.

In 1985, the Thai Cabinet approved a draft National Forest Policy that reaffirmed the standing policy of maintaining at least 40% forest cover in the country. Nature conservation in all fields was strengthened when the 1975 Enhancement and Conservation of National Environmental Quality Act was passed. This provided for the establishment of the National Environment Board within the Office of the Prime Minister, charged with submitting recommendations on environmental policy to the Cabinet (IUCN, 1991).

6 Forest management practices

Although forest fires in the tropics have been occurring since time immemorial, it is only now that serious attention is being given to them, primarily because of their devastating effects on life, properties and the environment. In the past, forest resources were plentiful and the adverse impacts of forest fires were considered to be insignificant. Now, in the wake of resource depletion and the degradation that threatens forest sustainability, the negative environmental impacts of forest fires in the South-east Asia region can no longer be ignored.

The occurrence of forest fires caused by natural phenomenon, e.g. lightning, is almost negligible in the region. Most of the documented occurrences of forest fires are of anthropogenic origin, either directly or indirectly aided by weather conditions and prolonged dryness (Fig. 8.1). Vast tracts of natural and plantation forests were burned annually, indicating that forest fire is a significant land management issue in the South-east Asia region (Haron and Manila, 1997).

Fig. 8.1. Oil palm plantations in Peninsular Malaysia.

Forest fire mismanagement is now one of the most serious problems facing forests in many parts of the world, especially in South-east Asia, and is rapidly becoming a global crisis. In 1997 alone, vast forest fires occurred in Indonesia and this was coupled with considerable numbers of fires in Malaysia and Vietnam. Most of the fires in the region were deliberately set, as they have been for many years, to clear forests for other purposes, particularly for agricultural plantations.

6.1 Brunei Darussalam

The problem of forest fires is not as severe as in some South-east Asian countries as most of the forest cover is maintained and comprises moist tropical rainforest. Sensitive forest types, such as mixed dipterocarp, montane and mangrove forests and peat swamps, are mostly found in remote areas far away from human settlements and activities. However, the dry forest types such as heath and coastal forests that are found in coastal areas are very susceptible to frequent fires because of long dry periods and windy conditions, high fuel contents and easy access. Once these areas are damaged by fires, further damage becomes an annual event.

The major source of bush fires is human activities, through either arson, agricultural practices or conversion of forest into other land uses. Wildfires usually start from roadsides and nearby agricultural areas, spreading through the nearby forest and scrub areas. An efficient fire prevention strategy therefore requires instilling awareness amongst the rural population with the use of multimedia information campaigns describing the devastating effects of fires on the environment. Another fire prevention measure that can be employed is the construction of firebreaks and fuelbreaks around and within the existing 30,000 ha of forest plantation in the country (Haron and Manila, 1997).

Currently, there is no specific legislation enforced to prevent and control fires, except for a provision in the Forest Law of 1984, Chapter 46, which prohibits slash-and-burn activities in reserved lands and details the penalties involved in such offences. The Fire Department is collaborating with relevant agencies in the country over the implementation of fire preventive measures by requiring the rural population, particularly farmers, to seek prior approval before burning agricultural residues over wide areas (Mahmud, 1996).

6.2 Indonesia

Forest areas in Indonesia that are prone to fire include deciduous types (particularly teak, *Tectona grandis*, forests), forest stands whose canopies have been opened up and/or broken, grasslands and secondary forests. Such areas are found in almost all the 27 provinces of the country.

Table 8.3 shows the importance of fire as an agent of land use and land class change in Indonesia (Haron and Manila, 1997). It further indicates that the major cause of these forest fires is anthropogenic activities. Even the 1982–1983, 1994 and 1997 fires in the Middle Mahakam area in East Kalimantan and Sumatra were associated with logging activities in the 1970s, coupled with transmigration and shifting cultivation (SOTAR, 1995).

The most serious outbreak of forest fire occurred in East Kalimantan during the period 1982–1983, when a total of 3.6 Mha of tropical forest was

Table 8.3. Average annual rate of forest conversion, 1982–1990.

Nature and/or factors influencing land use/land class changes	Average area (ha year^{-1})	Percentage
Development of crops such as rubber and oil palm	160,000	12.2
Transmigration and related infrastructure	300,000	22.8
Shifting cultivation (slash-and-burn agriculture)	300,000	22.8
Fires in the forest	100,000	7.6
East Kalimantan fires of 1982–1983 averaged for the period 1982–1990	378,000	28.7
Other factors such as spontaneous transmigration, illegal logging, mining and urban development	1,577,000	5.9
Total	2,815,000	100.0

burned (Haron and Manila, 1997). In 1994, another 161,800 ha of forested area were destroyed by fire. This devastating fire was caused by a long drought associated with the El Niño phenomenon and the burning of coal seams under the forests. The Ministry of Forestry is now intensifying forestry extension services and issues guidelines on prescribed burning as a fire prevention strategy for rural populations. It has also developed a National Forest Fire Plan indicating the networks/linkages of different agencies and organizations within and outside the country, including a forest fire danger rating system developed for 27 provinces, a response plan and a training component for those involved (Handadhari *et al.*, 1996). Indonesia has also set up a National Coordination Team on Land and Forest Fire Management that operates at the central, provincial and district levels.

6.3 Malaysia

In Peninsular Malaysia, there have been no major forest fires documented in natural forests, except for isolated outbreaks occurring in forest plantations where lands have been cleared for commercial monocultures. Records indicate that a cumulative total of 1100 ha of pine and *Acacia mangium* plantations was burned in the period of 1985–1995 during prolonged annual dry spells between the months of January–March, and June–August (Yunus and Yahaya, 1996). However, such fires have usually involved less than 100 ha and were readily brought under control.

Most of the documented occurrences of fire reveal that they are human-induced, caused by negligence or agricultural activities. The Forestry Department regularly undertakes forestry extension services and organizes training courses and fire drills among its staff as part of its fire prevention strategies. It has also created a National Forest and Plantation Fire Committee (NFPFC), with representatives from relevant government agencies to implement and monitor the Regional Response Plan prepared by the Department.

Meanwhile, in Sabah the threat of fire to the forest has been serious, and records show that about 1.0 Mha of mostly secondary forests were burned during the period 1983–1985 (SOTAR, 1995). The fires were attributed to a severe drought, which was also blamed in part for the forest fires in East Kalimantan. Most of the documented forest fires are caused by people, through either carelessness, negligence or misuse. The Forestry Department of Sabah is undertaking appropriate fire prevention and control strategies under the Forest Enactment of 1968, such as forestry extension services and training programmes to combat forest fires.

Forest fire incidents in Sarawak are confined to plantations, with minimal occurrences in the natural forest. To date, only relatively small plantations of *Acacia mangium* and *Shorea macrophylla* have been destroyed by fire (in

1991–1994), with the causes being related to agricultural activities in adjoining farm lands (SOTAR, 1995).

The fire prevention strategies followed by the Forestry Department of Sarawak are undertaken through forestry extension services directed towards the rural population and continuous training activities, including forest fire courses, among its staff. A guidelines/manual for fire protection in forest plantations has also been formulated by the Reforestation Unit of the Forestry Department.

6.4 Philippines

In the Philippines, human-induced fires account for large areas of degraded forest, in both natural and plantation areas. Destruction of forests by fire is further compounded by illegal harvesting and hillside farming by upland communities. The Forest Fire Control and Management Programme (FFCMP) that started in 1984 is presently handled by the Department of Environment and Natural Resources (DENR), covering seven fire-prone regions of the country.

6.5 Thailand

The forests of Thailand are categorized into two main groups: evergreen forest (45%) and deciduous forest (55%). The deciduous forests are further classified into three main subgroups: mixed deciduous forest, dry dipterocarp forest and savannah. As they shed their leaves during the dry season (December–April), they create high fuel loads and are extremely vulnerable to fire (Samran and Akaakara, 1996).

The Royal Forest Department of Thailand has a Forest Fire Control Division that directs the operations of four national forest fire control centres and 59 forest fire control stations distributed nationwide. Thailand is the first South-east Asian country in which helicopters and fixed-wing aircraft have been used for aerial fire-fighting and personnel transport on a regular basis (Haron and Manila, 1997).

7 Factors contributing to the haze episodes

Tropical rainforests have generally been regarded as ecosystems in which natural fire was excluded by fuel characteristics and the prevailing moist environment (Goldammer and Seibert, 1990). However, recent findings illustrate that climatic conditions since the late Pleistocene have favoured the occurrence of natural and anthropogenic fires in the Amazon Basin and in

South-east Asia. Since 1982, wildfires in the ASEAN region have destroyed more than 5 Mha of primary and secondary rainforest. These fires were the result of extensive land-clearing activities, and an extreme drought that was attributed to the El Niño–Southern Oscillation complex. Generally, heavily disturbed forests have suffered very serious fire damage, with shifting cultivation being implicated in the process of forest clearance and in increasing fire hazards. There has been speculation about the causes of drought, but some of the periodic dry periods in such places as eastern Borneo have been linked to oceanic and atmospheric fluctuations in the Pacific known as the 'Southern Oscillation', and particularly extreme anomalies in ocean-warming process and changes in currents in the equatorial regions of the eastern Pacific known as 'El Niño'. The link between the oscillation and its extremes is usually referred to as the El Niño–Southern Oscillation, or ENSO.

It is difficult to link forest clearance to these climatic changes directly, but the removal of vegetation cover has been related to forest fires in at least two indirect ways. Firstly, forest destruction may play a role in the intensity of El Niño because of the increased turbidity in the seas surrounding heavily logged islands such as Borneo. Large amounts of eroded silt and debris from cleared land affect water movements, sea surface temperatures and the air currents above them. It is clear that ENSO is not uniform in its patterns and effects in South-east Asia. Secondly, logged-over forests do not retain moisture as well as primary forest, and they dry out very quickly. An additional factor is that opening up forests produces trails and dead-wood residues, resulting in the accumulation of plant biomass close to and on the ground. Logging trails can act as wind tunnels and highways for fire, thus encouraging its spread (King, 1996).

As a result of forest fires, trace gases and particulate emissions in the form of thick smog are released into the atmosphere and, as a result of the simultaneous occurrence of an ENSO event, this smog is transported hundreds of kilometres to various cities in the region (Haron and Manila, 1997). The haze, resulting from the atmospheric concentrations of particulates, is not only a health hazard but has also caused disruption in the aviation and marine industries and to agricultural outputs.

The peat-swamp forests found in the lowlands of Borneo and some other parts of the ASEAN region represent another fuel type. With an increasing precipitation deficit and lowering of the water table in the peat swamp biome, the organic layers progressively dry out. According to Johnson (1984), as cited by Goldammer and Seibert (1990), during the 1982–1983 ENSO event, various observations in East Kalimantan confirmed that the swamps had dried out to depths of more than 1–2 m. Burning this type of forest usually produces thick smog as a result of the incomplete combustion that occurs underground. The intense smoke produced during the 1997 haze episode came from the burning peat bogs in Central Kalimantan and the Riau area of Sumatra. It was the burning of these bogs rather than the overall extent of fires that led to the intense smoke that engulfed several countries in the region.

As the population increases, the quest for land increases proportionately. As mentioned earlier, the population in the region is increasing rapidly, and the response of governments in the region has been the development of large-scale programmes tailored to meet their policies, such as transmigration, large-scale agricultural developments and forest plantations, and the development of new settlement areas. All these programmes entail the utilization and development of extensive forested areas (Sage, 1996). The quickest and cheapest tool for clearing the forests is fire.

Populations that are either attracted to or forced to migrate to tropical forest areas tend to adopt a non-sustainable slash-and-burn pioneer form of agriculture, without the long fallows of traditional systems. Both shifting cultivation and pioneer farmers depend on burning to produce crops at acceptable labour input intensities. As a result, huge areas are slashed and burned during a specific time period, producing vast amounts of smoke. Shifting cultivators destroy substantial amounts of valuable timber when clearing primary forest, and the clearing and burning of forest for cultivation cause long-term environmental deterioration in the tropical forest region. Shifting cultivation probably represents the single greatest threat to the integrity of Borneo's natural resources and is resulting in an unacceptable degree of human suffering (King, 1996).

8 Impacts of haze on tropical forests

Fires in the forest and other vegetation of the tropics and subtropics are having an increasing regional and global impact on the environment. Fire changes the physical state of the vegetation, releasing a variety of greenhouse gases into the atmosphere, with resultant negative implications for global warming. The release of chemically reactive gases during biomass burning strongly influences chemical processes within the troposphere. The smoke plumes from tropical biomass fires carry vast amounts of atmospheric pollutants, including carbon dioxide (CO_2), carbon monoxide (CO), nitrogen oxides (NO_x), nitrous oxide (N_2O), methane (CH_4), sulphate (SO_4), non-methane hydrocarbons, and aerosols. These pollutants from biomass burning are injected into differing strata of the atmosphere, depending on the rate of heat release, size of the fire, and the stability and wind velocity profile of the atmosphere.

The effects caused by air pollutants on forest ecosystems are related to the functional ability of forests to act as carbon sinks. Air pollutant emissions into the atmosphere cause tropospheric pollution and the trends in air pollution emissions are affected by the level and composition of economic activities of a nation, demographic influences, meteorological conditions, and regulatory efforts to control emissions (Haron and Manila, 1997). Emissions grow as the economy and population increase. In Malaysia alone, air pollutant emissions, mainly comprising CO_2 and CO, have been estimated to be about

850,000 t year^{-1}, an increase of 40% from 1977. The increase can be attributed to economic growth, with the transportation sector contributing 58% of the emissions (Sham *et al.*, 1991).

Peat-swamp forests found in the lowlands of Borneo and some other parts of the ASEAN countries represent another problem. While the spread of surface and ground fires in this type of organic terrain is not severe, deep burning of organic matter leads to the toppling of trees and the complete removal of standing biomass. As a result, most of the species diversity in the affected areas, including both flora and fauna, is likely to be lost. More than 5 Mha of forest and other lands have been destroyed in the region, with major losses to biodiversity, non-timber forest products and soil fertility; insects and wildlife, including many endangered species, suffered catastrophic decline. The Malesian district of South-east Asia alone is thought to harbour over 28,000 species of vascular plants (Ogino, 1991). Ogino (1991) estimated the number of invertebrates inhabiting the forest canopies in this region to be between 2.5 million and 30 million species, with the functions and interrelationships of the organisms within the canopy itself still being completely unknown. The forest ecosystem everywhere in South-east Asia has already deteriorated mainly because of human activities of various intensities, including forest fire.

The problem of acidic rain is not yet as serious in this region as in some countries of Europe and North America. The two major contributors to acidic rain are sulphur dioxide and nitrogen oxides, released into the atmosphere through industrial processes, motor vehicles, power generation and other human activities, including forest fires. With growing industrialization in South-east Asia, acidic deposition could also present a problem for forests if no proper preventive measures are taken.

The feedback mechanisms associated with anthropogenic forest degradation and climate change may also lead to changing precipitation regimes. These effects will probably be most dramatic in today's rainforest biomes, which will develop toward pyrophytic and xerophytic communities characterized by fires with short-return intervals and will further contribute to the changes in the atmosphere. Many tropical rainforests in the lowlands, even if untouched by human disturbance, would probably not survive a generally warmer and drier climate. Thus, the equatorial rainforest could migrate or be confined to small refugia at higher elevations (Fosberg *et al.*, 1990).

Global change in South-east Asia will substantially increase stresses on natural forests, but much more so on man-made forests, agricultural plantations, fields and the urban ecosystem (Fig. 8.2). In forests and agricultural plantation crops, there will be generally increased levels of physiological and ecological stresses and consequently instabilities, loss of self-regulatory capability and possibly reduced net primary productivity and less organic matter accumulation.

Fig. 8.2. View of Chiang Mai, in northern Thailand. As is apparent from the photograph, this city is already experiencing air quality problems.

Climate change, induced by the greenhouse effect, is expected to take place within the next 100 years – a time span comparable to the planting-to-harvest interval of many commercial tree species. The greenhouse theory is based on the energy balance between incoming solar energy and energy radiated to space from the earth (Houghton *et al.*, 1996). If there is no balance between the incoming and outgoing energy, then the earth will either warm or cool. The predicted increases of temperature are expected to be comparable to those that have taken place since the end of the last ice age 12,000 years ago. The rate of temperature change predicted from the increase in greenhouse gases, namely 5°C in 100 years, is believed to be unprecedented (Fosberg *et al.*, 1990).

In economic terms, the 1997 haze episode contributed to a loss of an estimated US$20 billion in the ASEAN region alone. Tourist bookings in the region fell by one-third. Ecotourism industries were particularly affected in the region, with major losses of revenue occurring. Such losses provide an important incentive to improve forest protection across the region.

9 Future challenges

Some key steps should be taken by the international community to rectify the problems described above. These include changes in management practices to minimize environmental risks. In particular, we recommend the introduction of standard fire prevention techniques, such as the creation of firebreaks, the prevention of gaps in the forest canopy, fire watching and fire prediction techniques, as well as fuel reduction techniques such as thinning from below and prescribed burning. In addition, there should be a wide-ranging series of educational activities to inform people about forest fire hazards.

Carefully managed educational promotions financed by governments, industry or non-governmental organizations (NGOs) can help to reduce the impact of forest degradation by providing guidelines for sound forest management. As an example, the International Tropical Timber Organization (ITTO) has developed the ITTO Guidelines on Fire Management in Tropical Forests, which were designed to provide a basis for policy makers and managers at various levels to develop programmes and projects related to fire in tropical natural and planted forests (ITTO, 1997).

The enforcement of existing laws is a prerequisite to any successful policy to curb people from practising hazardous activities in the forest ecosystem that could cause detrimental effects to the environment. For instance, most of the 1997 fires were set by farmers and companies wishing to clear their land for agricultural or plantation purposes. As a result of the severe drought caused by the El Niño weather pattern and the lack of knowledge on fire prevention and control, fires rapidly developed out of control and razed hundreds of thousand of hectares of forest in the region.

Unchecked land, bush and forest fires are threatening to become a disaster of regional or even global proportions. However, simple and inexpensive changes to forestry practices could greatly reduce fire risks. An example would be to introduce a licensing system under which large companies would have to obtain a permit before using fire to clear land. There would have to be additional restrictions during El Niño years and perhaps a permanent ban on burning in peat bogs. The governments of South-east Asian countries could also tackle the structural features of the economy that encourage fires. For instance, if timber prices were allowed to rise it would be much more attractive for plantation companies to harvest forest timber than to burn it.

According to Brown *et al.* (1996), cited in Abdul Rahim and Abdul Razak (1996), forest management practices to curb CO_2 increases should be grouped into three categories: (i) management for carbon conservation; (ii) management for carbon storage; and (iii) management for carbon substitution. Management for carbon conservation entails preventing carbon emissions by conserving existing carbon pools in forest vegetation and soil as much as possible through options such as controlling deforestation, changing harvesting regimes, and controlling other anthropogenic disturbances such as fire and pest outbreaks. The goal of storage management is to expand the storage of carbon in forest ecosystems by increasing the area and/or biomass of vegetation, increasing soil carbon density by preserving natural and plantation forests, and increasing storage in durable wood products. Substitution management aims at increasing the transfer of forest biomass carbon into products (e.g. construction materials and biofuels) rather than using fossil fuel-based energy products, cement-based products, and other non-wood building materials.

Forests are a major component of the global carbon cycle: they account for 80% of annual exchanges of CO_2 between the land and atmosphere, and may absorb up to 25% of the 6 billion t of carbon currently emitted each year by the combustion of fossil fuels. However, deforestation also causes the emission of around 1.7 billion t of carbon per year. Tropical forests could help to reduce the rate of climate change both by conserving existing stocks of carbon in forests that are currently being lost, and by creating new stocks of carbon in growing trees, i.e. through forest plantations.

For plantations to be successful in storing carbon, they need to be worked over a long rotation period (a minimum of 50 years) and serious consideration needs to be given to the best end-use strategies for this new plantation timber. Long-lasting wood products such as furniture or wooden buildings would be more effective at keeping carbon out of the atmosphere than short-life disposable products such as paper.

Forest plantations alone cannot stop the accumulation of carbon dioxide in the atmosphere, although they have sometimes been proposed as an important measure to combat climate change. According to the World Resources Institute, as cited by Luukkanen (1991), annual global CO_2 emissions comprise 7 billion t of carbon (of which 1.7 billion t or 25%, is of biotic origin, mainly from tropical deforestation). In order to absorb this amount of CO_2, 1–1.5 billion ha of new tropical forests would be needed. If these forests were to be planted over a period of 50 years, 20–30 Mha of new forest would be needed annually, something that is very unlikely to happen – the current rate of reforestation is about 1 Mha annually (Goh, 1991). The establishment and maintenance costs of any such operation would be very high. A Beijer Institute and UN Environment Programme (UNEP) study, cited by Goh (1991), calculated the cost of 500 Mha of tropical forest plantation at between US$200 and 400 billion.

An alternative strategy would be to restrict and eventually eliminate shifting agriculture. In some cases, farmers would have to be resettled and alternative forms of land use such as plantation cultivation and commercial forestry would be encouraged. There is increasing evidence that consumers would like future timber supplies to come from sustainably managed timber plantations rather than from secondary or primary forests.

In achieving a better and safer environment, the industrialized nations agreed at the Kyoto summit in 1997 to cut emissions of carbon dioxide and other gases by an average 5.2% between 2008 and 2012, with a 7.0% cut for the USA and 8.0% for the European Union (Oberthür and Ott, 1999). The same initiatives need to be taken immediately by South-east Asia countries in reducing the amount of carbon dioxide and other gases being released into the atmosphere, and which are giving rise to significant environmental problems in the region.

10 Conclusions

Human impacts on tropical forest lands in South-east Asia are rapidly increasing, causing overall degradation, and conversion of rainforest vegetation to pyrophytic life forms with increased flammability and fire frequency. Furthermore, increasing populations are resulting in cultivation extending from plains and valleys into hill areas. As a result, an estimated 160 Mha of forested land in the region have been seriously degraded by fires of anthropogenic origin (Mercedes, 1993).

Much could be done immediately to translate what we already know into action. Steps include reforming the policies of international lending institutions and development assistance programmes to give greater consideration to the environmental impacts of policies that either provoke or eliminate fires. For instance, the recent increase in the environmental review capabilities of the World Bank is a hopeful sign, but it is only a very small beginning.

Education must now begin to bring about long-term changes in attitudes towards fire and nature. Related agencies must promote and facilitate the development and implementation of educational and public awareness programmes on climate change and its effects. At the same time, global fire weather monitoring systems must be expanded and coordinated. Meanwhile, the governments of countries in the South-east Asia region must promote the development and implementation of education and training programmes, including the strengthening of national institutions and the exchange or secondment of personnel to train experts in this field.

The issue of transboundary air pollution is related to the twin themes of population growth and increasing industrialization. As we grow together as a region, the threat from a source originating from beyond our national borders can become significant if not contained early enough. The solution is to have transboundary consultation, cooperation and approaches. In this region, we are fortunate because the ASEAN forum and many bilateral arrangements provide the available forums for such activities. More importantly, the will to have such cooperation exists and indications that ASEAN member countries are responding very seriously to the threat are evident through events and agreements such as the following:

1. The Kuala Lumpur Accord on Environment and Development, adopted by the ASEAN Ministers for the Environment in June 1990, which reflected ASEAN's sense of urgency by calling *inter alia* for efforts leading towards 'the harmonization of transboundary pollution and abatement practices'.
2. The informal ASEAN Ministerial Meeting on the Environment held in Kuching, Malaysia, on 21 October 1994, whereby the ministers agreed to enhance cooperation to manage natural resources and control transboundary pollution within the ASEAN region, to develop a regional early warning system, and to improve the capability of member countries in these areas.

3. The 'ASEAN Cooperation Plan on Transboundary Pollution' that was formulated in June 1995 in Kuala Lumpur. Thereby, all ASEAN member countries agreed to collaborate actively towards developing expertise and capability to minimize the effects of transboundary pollution with assistance from some donor countries.

To set goals for a more haze- and pollutant-free environment, the crux of the issue is in the retention of adequate forest cover for all countries in the region. This rather transparent goal, if attained, would ensure that the system could cope with large-scale disasters and that it can continue to maintain other forest goods, including the provision of forest products, the conservation of soil, water and biological resources, the mitigation of climate change, and the maintenance of environmental stability and an ecological balance.

Pollution, including its transboundary variant, is the process of overloading the earth's ecosystems with damaging materials or waste energy. It has grown from a local nuisance to a national and even global menace. Action has to be taken now by governments, municipalities and industries in both higher and lower income countries. Whilst the magnitude of the challenges is daunting, the prospects are there for real progress, especially in the regional context. Both regulatory and economic instruments should be mobilized and fully utilized.

References

Abdul Rahim, N. and Abdul Razak, M.A. (1996) Carbon conservation and sequestration in tropical forests. Research status and perspectives. In: Haron, A.H., Dahlan, T., M. Puat, D. and Amran, M. (eds) *Transboundary Pollution and the Sustainability of Tropical Forests*. ASEAN Institute of Forest Management, Kuala Lumpur, Malaysia, pp. 333–348.

Aksorn, S. (1991) Welcoming address. In: Kongtong, P., Bhumibhamon, S., Wood, H. and Bonnpragob, K. (eds) *Proceedings of the National Biological Conference of Thailand, Including Asia and the Pacific Regions, on Global Change: Effects on Tropical Forests, Agricultural, Urban and Industrial Ecosystems, Bangkok, 22–24 October, 1990*. Ramkamhaeng University, Bangkok, pp. 101–108.

Bautista, G.M. (1990) The forestry crisis in the Philippines: nature, causes and issues. *The Developing Economies*, 28 March 1990, pp. 67–94.

Beang, L.C. (1996) Community forestry in Cambodia. In: *Proceedings of the International Conference on Community Forestry as a Strategy for Sustainable Forest Management*. Department of Environment and Natural Resources, Manila, Philippines, pp. 347–350.

Brown, S., Sathaye, J., Cannell, M. and Kauppi, P.E. (1996) Management of forests for mitigation of greenhouse gas emissions. In: Watson, R.T., Zinyowera, M.C., Moss, R.H. and Dokken, D.J. (eds) *Climate Change 1995. Impacts, Adaptations and Mitigation of Climate Change: Scientific–Technical Analyses*. Cambridge University Press, Cambridge, pp. 773–797.

Cameron, O. (1996) Japan and South-East Asia's environment. In: Parnwell, M.J.G. and Bryant, R.L. (eds) *Environmental Change in South-East Asia: People, Politics and Sustainable Development*. Routledge, New York, pp. 67–94.

Celso, R.R. (1986) Factors influencing environmental policies in developing countries. Paper presented at the Regional Symposium on Environmental and Natural Resources Planning, Manila, 19–21 February 1986.

Chudchawan, S. (1996) Community forestry programmes in Thailand: past, present and future. In: *Proceedings of the International Conference on Community Forestry as a Strategy for Sustainable Forest Management*. Department of Environment and Resources, Manila, Philippines, pp. 235–248.

Cuc, L.T. (1995) Forest ecosystem and swidden agriculture in uplands of northern Vietnam. In: Sorensen. K.W., Enriques, G.L., Umaly, R.C. and Kartana, J.T. (eds) *Proceedings of the Regional Seminar-Workshop on Tropical Forest Ecosystems Research, Conservation and Repatriation*. SEAMEO BIOTROP South-East Asian Regional Centre for Tropical Biology, Bogor, Indonesia, pp. 269–289.

FAO (1992) *The Forest Resources of the Tropical Zone by Main Ecological Regions*. UNCED, Rio de Janeiro, Brazil, June 1992.

Farmer, A.S.D., Caldecott, J.O., Philips, A., Prince, G. and Thomson, N. (1986) *Negara Brunei Darussalam Masterplan. Special Report: Wildlife Conservation and Management*. Huszar Brammah and Associates/Department of Town and Country Planning, Bandar Seri Begawan.

Fosberg, M.A., Goldammer, J.G., Rind, D. and Price, C. (1990) Global change: effects on forest ecosystems and wildfire severity. In: Goldammer, J.G. (ed.) *Fire in the Tropical Biota: Ecosystem Processes and Global Challenges*. Springer-Verlag, Berlin, pp. 463–484.

Goh, K.S. (1991) Action by global communities to reduce global environment change. In: Kongtong, P., Bhumibhamon, S., Wood, S. and Boonpragob, K. (eds) *Proceedings of the National Biological Conference of Thailand, Including Asia and the Pacific Regions, on Global Change: Effects on Tropical Forests, Agricultural, Urban and Industrial Ecosystems, Bangkok, 22–24 October, 1990*. Ramkamhaeng University, Bangkok, pp. 101–108.

Goldammer, J.G. and Seibert, B. (1990) The impacts of droughts and forest fires on tropical lowland rain forest of East Kalimantan. In: Goldammer, J.G. (ed.) *Fire in the Tropical Biota: Ecosystem Processes and Global Challenges*. Springer-Verlag, Berlin, pp. 13–31.

Handadhari, T. and Justianto, A. (1996) Forest fire management in Indonesia. Country report on the AIFM Forest Operations Technical Working Group (TGW) Meeting, Chiangmai, Thailand, 14 pp.

Haron, A.H. and Manila, A.C. (1997) Forest fire management and its influence on forest development in ASEAN. In: Haron, A.H., Dahlan, T., M. Puat, D. and Amran, M. (eds) *Proceedings of the AIFM International Conference on Transboundary Pollution and the Sustainability of Tropical Forests: Towards Wise Forest Fire Management*. ASEAN Institute of Forest Management, Kuala Lumpur, pp. 81–91.

Holdgate, M. (1994) Sustainability in the forest. In: Chiew, T.H., Ahmad, S.S., Tsai, L.M., Nor Aini, A.S. and Ahmad Zainal, M.I. (eds) *People, the Environment and Forestry – Conflict or Harmony*. Proceedings of the 14th Commonwealth Forestry Conference, Kuala Lumpur, 1993. Ministry of Primary Industries, Kuala Lumpur, pp. 23–33.

Houghton, J.T., Meira Filho, L.G., Callander, B.A., Harris, N., Kattenberg, A. and Maskell, K. (eds) (1996) *Climate Change 1995. The Science of Climate Change.* Cambridge University Press, Cambridge.

Ishemat, S. and Lemmens, R.H.M.J. (1994) *Plant Resources of South-East Asia.* No. 5(1). Prosea, Bogor, Indonesia.

ITTO (1997) *ITTO Guidelines on Fire Management in Tropical Forests.* ITTO Policy Development Series No.6. International Tropical Timber Organization (ITTO), Yokohama, Japan.

IUCN (1991) *Protected Areas of the World – a Review of National Systems,* Vol. 1. IUCN, Gland, Switzerland.

Javier, J.A. (1996) Accomplishment, experiences gained and goals in sustainable tropical forest management in the Philippines. In: *Proceedings of the ASEAN Seminar on Sustainable Tropical Forest Management: Progress Towards Achievement of the ITTO Year 2000 Objective.* Ministry of Forestry, Republic of Indonesia, Jakarta, pp. 115–138.

Johnson, B. (1984) *The Great Fire of Borneo. Report of a Visit to Kalimantan Timur.* WWF, Gland, Switzerland.

Kamis, A. and Mohd. Basri, H. (1996) Community forestry in Malaysia: overview, constraints and prospects. In: *Proceedings of the International Conference on Community Forestry as a Strategy for Sustainable Forest Management.* Department of Environment and Natural Resources, Manila, Philippines, pp. 175–190.

King, V.T. (1996) Environmental change in Malaysian Borneo: fire, drought and rain. In: Parnwell, M.J.G. and Bryant, R.L. (eds) *Environmental Change in South-East Asia. People, Politics and Sustainable Development.* Routledge, New York, pp. 165–189.

Luukkanen, O. (1991) Global change and the eco-physiology of tropical trees. In: Kongtong, P., Bhumibhamon, S., Wood, H. and Boonpragob, K. (eds) *Proceedings of the National Biological Conference of Thailand, including Asia and the Pacific Regions, on Global Change: Effects on Tropical Forests, Agricultural, Urban and Industrial Ecosystems, Bangkok, 22–24 October, 1990.* Ramkamhaeng University, Bangkok, pp. 66–72.

Mahmud, Y. (1996) *Forest Fire Management in Brunei Darussalam.* Country Report presented to the AIFM Forest Operations Technical Working Group (TWG) Meeting, Chiangmai, Thailand.

Mercedes, B.C. (1993) Population growth in South-East Asia: pushing the limits. In: Brookfield, H. and Byron, Y. (eds) *South-East Asia's Environmental Future: the Search for Sustainability.* Oxford University Press, Kuala Lumpur, Malaysia, pp. 33–45.

Mursidin, A.S. (1996) Social/community forestry development in Indonesia. In: *Proceedings of the International Conference on Community Forestry as a Strategy for Sustainable Forest Management.* Department of Environment and Natural Resources, Manila, Philippines, pp. 111–121.

Nguyen, T.T. and Le, H.T. (1996) Country report: Vietnam. In: *Proceedings of the ASEAN Seminar on Sustainable Tropical Forest Management: Progress Towards Achievement of the ITTO Year 2000 Objective.* Ministry of Forestry, Republic of Indonesia, Jakarta, pp. 145–167.

Oberthür, S. and Ott, H.E. (1999) *The Kyoto Protocol. International Climate Policy for the 21st Century.* Springer Verlag, Berlin.

Ogino, K. (1991) Urbanisation and land planning: causes and effects of climatic change. In: Kongtong, P., Bhumibhamon, S., Wood, H. and Boonpragob, K. (eds) *Proceedings of the National Biological Conference of Thailand, including Asia and the Pacific Regions, on Global Change: Effects on Tropical Forests, Agricultural, Urban and Industrial Ecosystems, Bangkok, 22–24 October, 1990.* Ramkamhaeng University, Bangkok, pp. 62–65.

Phengdouang V. (1995) Natural forest structure and regeneration after logging in Lao PDR and future management factors. In: Sorensen, K.W., Enriques, G.L., Umaly, R.C. and Kartana, J.T. (eds) *Proceedings of the Regional Seminar-Workshop on Tropical Forest Ecosystems Research, Conservation and Repatriation.* SEAMEO BIOTROP South-East Asian Regional Centre for Tropical Biology, Bogor, Indonesia, pp. 221–231.

Sage, C.L. (1996) The search for sustainable livelihoods in Indonesian transmigration settlements. In: Parnwell, M.J.G. and Bryant, R.L. (eds) *Environmental Change in South-East Asia: People, Politics and Sustainable Development.* Routledge, New York, pp. 97–122.

Saidin, S. (1996) Country report: Brunei Darussalam. In: *Proceedings of the ASEAN Seminar on Sustainable Tropical Forest Management: Progress Towards Achievement of the ITTO Year 2000 Objective.* Ministry of Forestry, Republic of Indonesia, Jakarta, pp. 62–73.

Samran, S. and Akaakara, S. (1996) *Forest Fire Management in Thailand.* Country Report to the AIFM Forest Operations Technical Working Group (TWG) Meeting, Chiang Mai, Thailand.

Sham, S., Cheang, B.K., Leong, C.P. and Lim, S.F. (1991) *The August 1990 Haze in Malaysia.* Technical Report No. 49, Malaysian Meteorological Service, Kuala Lumpur.

SOTAR (1995) *State-of-the-Art-Review Report on Forest Fire Management in ASEAN.* ASEAN Institute of Forest Management, Kuala Lumpur, Malaysia.

U Than, S. (1996) Community forestry in Myanmar. In: *Proceedings of the International Conference on Community Forestry as a Strategy for Sustainable Forest Management.* Department of Environment and Natural Resources, Manila, Philippines, pp. 351–361.

Yunus, Z. and Yahaya, M. (1996) *Forest Fire Management in Malaysia.* Country Report to the AIFM Forest Operations Technical Working Group (TWG) Meeting, Chiang Mai, Thailand.

Research on Air Pollution Impacts on Indian Forests 9

M. Agrawal[1] and S.B. Agrawal[2]

[1]*Department of Botany, Banaras Hindu University, Varanasi, India;*
[2]*Department of Biological Sciences, Allahabad Agricultural Institute, Allahabad, India*

Rapid technological and industrial advancement coupled with increases in population growth and urban expansions have triggered the deterioration of air quality in India. The problem of air pollution has assumed serious proportions in the metropolitan cities and around industrial areas. Research on air pollution impacts on forests mostly has been conducted around large point sources of pollution. The analysis of air quality data clearly shows SO_2, NO_2 and suspended particulate matter (SPM) as major pollutants of concern. SPM concentration was found to be above the standards set by the Central Pollution Control Board (CPCB) at most of the monitoring stations throughout the country. SO_2 and NO_2 concentrations are highest in the northern region and lowest in the southern part of the country. High concentrations of the secondary pollutant O_3 have also been reported around some cities.

Studies conducted downwind of industrial sources have clearly shown the negative impact of air pollutants on stem diameter, leaf weight, flowering and fruiting of trees. Visible injury symptoms have also been observed on tree leaves near industrial areas. Reductions in chlorophyll and ascorbic acid contents, percentage photosynthetic area, leaf weight and concentrations of N and P were common observations in plants growing in polluted areas. In contrast, foliar SO_4^{2-}-S and trace elements were higher in trees growing closer to polluting sources. Foliar SO_4^{2-}-S/organic S ratio was suggested to be a useful indicator of state of forest trees under S pollution stress. Air pollution has also been found to decrease the number of individuals and species richness of forests; the changes in

 Air Pollution and the Forests of Developing and Rapidly Industrializing Regions (eds J.L. Innes and A.H. Haron)

species diversity increase the proportion of resistant species in the community. Litter decomposition rate was found to be reduced at polluted sites. Nutrient release was also low in F-contaminated litter compared to F-free litter. In general there are very few exposure studies on tree species conducted at known concentrations of pollution. These studies have also confirmed that SO_2 has an adverse effect on trees on long-term exposure.

The existing data clearly show that air pollution in India has great potential for affecting the forest vegetation negatively. Most of the forests in the country have not been studied in relation to air pollution stress. Therefore, a national programme is warranted to identify the areas of forests with varying risk potentials in relation to air pollution.

1 Introduction

The significant contribution of air pollution to problems such as human health, the loss of agricultural productivity and forest decline has been a cause of increasing scientific and public concern throughout the world. Rapid technological, industrial and agricultural advances coupled with increases in population growth and urban expansion have triggered the deterioration of environmental quality. By the mid-1950s, most cities in Europe, the USA and Japan were experiencing severe air pollution problems as a result of the post-Second World War economic boom. However, similar problems are now affecting the developing world, including India, which is witnessing rapid industrial and urban development.

India, at present, has one of the fastest developing economies of the world. In terms of gross output, it ranks amongst the ten top industrialized nations. In parallel with the economic transformation, the region has also experienced strong population growth, from 342 million in 1947 to more than 900 million in 1994, accompanied by a concentration of the increasing population in urban areas. The problem of air pollution has assumed serious proportions in the metropolitan cities because of the concentration of industrial and tertiary activities around these centres. Air pollution is steadily increasing because of spiralling fossil fuel consumption by large fossil-fuelled electricity generating plants and the transport sector. Fossil fuel consumption has increased from 75 Mt year^{-1} in 1964 to 245 Mt year^{-1} in 1990 (Shrestha *et al.*, 1996). The number of vehicles has increased from 1.86 million in 1971 to 32 million in 1996 and is expected to increase further to 53 million by 2000 (Varshney *et al.*, 1997). While industrial air pollution is localized, mobile sources have emerged as the most significant contributor to regional air pollution.

In contrast to most developed countries, where air pollution research has shifted from the impacts of coal smoke and its associated pollutants to the effects of regional pollution on vegetation and human health, research in developing countries is mostly concentrated around large point sources of pollutants. In India, air pollution impacts are evaluated in relation to both human health and vegetation effects. However, air quality standards are largely based on human health impacts. The deterioration in air quality has adversely affected human health, vegetation and monuments. The recent emphasis on the economic liberalization in the country is likely to aggravate air pollution problems in the 21st century.

Publications relating to air pollution impacts on vegetation have largely focused on the effects of pollutants from industrial sources on agricultural crops; information on the effects of air pollutants on the forests of India is still very limited. This review examines the information dealing with air pollution impacts on Indian forests.

2 Forest types of India

India's forest cover was estimated to be 63.34 Mha in 1997, accounting for 23% of the geographical area (Anon., 1997a). The National Forest Policy of 1952, however, recommended that 33% of the land area should be under forest. While the natural forest cover has decreased, the area under plantations has increased considerably. The loss of natural dense forests is a cause of great concern for Indian foresters. At the national level, deforestation caused by conversion of forest land to development projects and urban expansion, over-exploitation of forest products for a variety of uses, forest fire and diseases, and increased use of plantations of exotic species are identified as the major causes of the decline in the natural forest cover in India. Overgrazing by cattle, inadequate fire protection and heavy firewood collection are the reasons for the very low regeneration of forests. The total dense forest area declined from 46.42 Mha in 1972–1975 to 26.73 Mha in 1993–1995 (Anon., 1997a). Several instances of forest pests and diseases have also been reported, such as the sal heartwood-borer (*Hoplocerambyx spinicornis*) in Madhya Pradesh (Anon., 1987), the deodar defoliator (*Ectropis deodarae*) in Himachal Pradesh (Anon., 1989) and the wilting of shishum (*Dalbergia sissoo*) in an area extending from the fertile Ganges plains to the Terai grasslands bordering Nepal (Anon., 1995b).

Puri *et al.* (1983) have given a comprehensive review of Indian forests. They are classified, on the basis of temperature, into: (i) tropical, usually up to 1000 m altitude; (ii) montane subtropical between 1000 and 1600 m altitude; (iii) temperate above 1600 m altitude on the mountains of Himalayas and Nilgiris; and (iv) alpine, from 3000 m to 4000 m altitude. Tropical forests have been further classified on the basis of their relative degrees of wetness into: (i) tropical wet evergreen forests on the western coast, north-east and Andamans islands receiving over 250 cm of annual rainfall; (ii) tropical moist semi-evergreen forests in northern Assam, Bengal and parts of Orissa receiving heavy rainfall; (iii) tropical moist deciduous forests in moist areas of Kerala, Karnataka, southern Madhya Pradesh, eastern Uttar Pradesh, Bihar, Bengal and Orissa; and (iv) tropical dry deciduous forests in drier parts of Uttar Pradesh, Bihar, Maharashtra, northern Madhya Pradesh, Tamil Nadu, Punjab and Andhra Pradesh.

Most of the industrial activities and urban centres are concentrated in the area dominated by tropical forests, hence these are the most at risk from the adverse effects of air pollutants.

3 Variations in air quality in India

The air pollutants of major concern in India are particulate and gaseous pollutants. Particulate matter includes cement dust, coal dust, soot, fly ash, heavy metal particles, road dust and other materials. Gaseous pollutants include SO_2, nitrogen oxides (NO_x), photochemical oxidants, hydrogen fluoride (HF), hydrogen sulphide (H_2S), chlorine and ammonia. The major air pollutants recognized in India are listed in Table 9.1, together with their emission sources. The areas referred to in the text are shown in Fig. 9.1.

Air quality monitoring began in India in the late 1960s, but only a few pollutants, namely SO_2, NO_x, suspended particulate matter (SPM) and lead, were monitored. A database on air quality in Indian cities representing a nationwide cross-section of different industrial, geographical and climatic conditions has been prepared by the monitoring network of the National Environmental Engineering Research Institute (NEERI), Nagpur, and the Central Pollution Control Board (CPCB). The analysis of NEERI air quality data for annual average SO_2 concentrations reveals a trend for increasing concentrations (from 10 to 40 $\mu g\ m^{-3}$) in most of the areas in northern regions, except for a few cities, including Delhi, that had mean annual SO_2 concentrations above 60 $\mu g\ m^{-3}$ after 1985 (Agarwal *et al.*, 1999). The industrial belts of the eastern and western regions of the country also have high mean annual SO_2 concentrations, ranging from 70 to 85 $\mu g\ m^{-3}$. In the southern region, the annual averages have always remained within the standard. In Delhi, total anthropogenic SO_2 emissions are approximately 59,000 t year^{-1}. Annual average concentrations of SO_2 ranging from 120 $\mu g\ m^{-3}$ to 66 $\mu g\ m^{-3}$ between 0.5 to 6 km, respectively, in the prevailing wind direction from a 1500 MW thermal power plant situated in Obra have been reported (Singh *et al.*, 1990; Pandey, 1993). Seasonal variations were quite significant for SO_2, with concentrations being highest in winter and lowest during the rainy season.

NO_x concentrations have increased dramatically since 1990 (Agarwal *et al.*, 1999). In Mumbai, Delhi and Chennai, traffic density contributed 52–72% of the total estimated NO_x in 1990. Annual average NO_x concentrations varied from 10 to 90 $\mu g\ m^{-3}$ in various parts of the country, being

Table 9.1. Sources of certain major and minor air pollutants.

Pollutants	Typical sources
Sulphur dioxide (SO_2)	Thermal power plants, petroleum industry, oil refinery, smelters, domestic use of fuel, sulphuric acid plants
Oxidants (O_3)	Produced during photochemical reaction involving primary pollutants, such as NO_x and hydrocarbons (emitted from combustion of fossil fuels from automobiles) in presence of sunlight
Hydrogen fluoride (HF)	Ore smelting, phosphate fertilizers, aluminium factories, ceramic and glass manufacturing industries, brick kilns
Nitrogen oxides (NO_x)	Combustion of fossil fuel, oil refining, explosives industry, automobile exhausts, N fertilizer plants
Ammonia (NH_3)	Fertilizer factories, accidental spills during manufacture of anhydrous ammonia
Chlorine (Cl_2)	Manufacture of HCl and plastics, accidental spills
Hydrogen sulphide (H_2S)	Viscose rayon plants, petroleum industry, sewage treatment, tanning industry, dye manufacture and oil refining
Particulates	Thermal power plants, transportation, building construction, stone crushing, manufacture of cement, iron and steel, smelters, coal combustion, forest product processing, agriculture, industrial activities, soil erosion

Fig. 9.1. Map of India showing locations of sites mentioned in the text.

highest at Gajroula in Uttar Pradesh. In the southern region, the annual mean concentration of NO_x was quite low, except for a few cities. Of the southern cities, Pondicherry recorded the highest NO_x annual average concentration, namely 64.1 $\mu g\ m^{-3}$. In many towns of Bihar, high NO_x levels have also been recorded. In Howarh (eastern region), NO_x annual mean concentrations were above the standard of 60 $\mu g\ m^{-3}$. In the western region, the annual mean concentration of NO_x has increased in the industrial towns of Surat, Vapi, Rajkot, and Ankleshwar in Gujarat, but is still below the standard. In Ahmedabad, however, an annual mean concentration of 490 $\mu g\ m^{-3}$ was recorded in 1991.

High NO_x concentrations have been recorded in small towns, such as in the hill towns of Himachal Pradesh, and Dehradun in Uttar Pradesh, Alwar in Rajasthan, and Agartala, the capital of Tripura. Even the towns in north-east India have shown increasing trends in NO_x concentrations. Annual mean concentration of NO_2 ranged from 19 to 59 $\mu g\ m^{-3}$ in different zones of Varanasi city, situated on the eastern Ganges plain of Uttar Pradesh. There, a peak concentration of 159 $\mu g\ m^{-3}$ was recorded during winter (Pandey *et al.*, 1992; Pandey and Agrawal, 1994).

Dust (SPM) pollution is emerging as the most serious problem in India. It is derived from both natural and combustion sources. In 1973, 67% of the total industrial SPM emissions were from power plants (NEERI, 1991b). However, the transport sector doubled its share of the total SPM emissions between 1970 and 1990. In Calcutta, a 60% increase in SPM emissions between 1970 and 1980 was recorded. Total SPM emissions in Delhi were estimated at around 115,700 t $year^{-1}$ in 1990, and this figure is projected to increase to 122,600 t $year^{-1}$ by 2000 (NEERI, 1991a). Natural dust from the Great Indian Desert also contributes to the Delhi SPM levels. Annual average SPM levels in Delhi were recorded as 400 $\mu g\ m^{-3}$ and peak levels have remained above 1000 $\mu g\ m^{-3}$ since 1987. The southern region of the country has lower levels of SPM than those recorded in the northern region. Pondicherry had the highest annual mean, with 181 $\mu g\ m^{-3}$ in 1994 (Anon., 1997b). SPM levels in West Bengal, Orissa and Bihar were higher than in the north-east of the country. In Calcutta, annual mean SPM concentrations were above 350 $\mu g\ m^{-3}$, with peak concentrations being above 1000 $\mu g\ m^{-3}$ since 1987 (Anon., 1995b). In the western region, big towns and industrial sites exceed critical levels for SPM; these are currently 140 $\mu g\ m^{-3}$ in rural and residential areas and 360 $\mu g\ m^{-3}$ in industrial areas. In Varanasi, annual average SPM levels varied between 126 $\mu g\ m^{-3}$ and 336 $\mu g\ m^{-3}$ between clean and the most polluted sites, respectively (Pandey *et al.*, 1992). Annual average SPM concentrations and dust fall levels downwind of a cement factory at Dala in the south-eastern part of Uttar Pradesh were 595 $\mu g\ m^{-3}$ and 7 $g\ m^{-2}\ day^{-1}$ and 100 $\mu g\ m^{-3}$ and 1 $g\ m^{-2}\ day^{-1}$ at heavily and the least polluted sites, respectively (Agrawal and Khanam, 1997).

As a result of the dramatic increase in the number of motor vehicles in Indian cities, atmospheric accumulations of the secondary air pollutant ozone (O_3) have also been reported from some parts of the country. Varshney and Agrawal (1992) found that ground-level O_3 concentrations were between 20 and 273 μg m^{-3} in Delhi. In another study, conducted by the Central Road Research Institute (CRRI) at seven sites in Delhi, the 8 h mean O_3 concentration during the day exceeded the WHO mean standard of 100–200 μg m^{-3} by 10–40% (Singh *et al.*, 1997). The maximum O_3 concentration reached 251 μg m^{-3} in an area of heavy traffic congestion. Annual average O_3 concentrations of 16–48 μg m^{-3} were reported in and around Varanasi city (Pandey and Agrawal, 1992). The highest O_3 concentration occurred in summer. In a diurnal cycle, peak concentrations of O_3 were recorded during early and late afternoon in winter and summer, respectively. Surface O_3 measurements for 1 year in 1991–92 at an urban site in Pune and at core zones of the Nilgiri Biosphere forests located in Tamil Nadu and Karnataka states in southern India revealed annual average O_3 concentrations of 27 ppb (54 μg m^{-3}) at Pune and 10-day average O_3 concentrations of 15 ppb (30 μg m^{-3}) at forested sites (Khemani *et al.*, 1995a). The diurnal variation of O_3 in the forest area was different from that at Pune.

Research efforts into precipitation chemistry have attempted to understand the role of SO_2 and NO_x emissions in the generation of an acidic deposition problem in India. Acidic deposition was first reported in 1974 at Chembur, a major industrial area in Mumbai city, Maharashtra State (Mahadevan *et al.*, 1984). Khemani *et al.* (1994) have analysed precipitation samples from 11 locations in an industrial area on the west coast of India and found the pH to vary between 5.87 and 7.05. The presence of excess cations, particularly Ca^{2+}, was the cause of the neutralization of acidity. Precipitation near industrial sources is normally acidic (pH 5.0 to 5.7) because of an excess of anions, especially SO_4^{2-} (Khemani *et al.*, 1989). Rainwater analysis at the Silent Valley forest in Kerala during the monsoon indicated an average pH of 5.31. The results of a study conducted to measure acidic deposition downwind of a super thermal power plant in northern India revealed pH values of 5.3 at 500 m, and 5.9 at 1 and 4 km from the source in the prevailing wind direction during the monsoon (Khemani *et al.*, 1995b). The study concluded that the formation of acidic deposition depends not only on the acidity present in the rain but also on the amount of alkaline matter present, which acts as a neutralizing agent.

Lead (Pb) is also recognized as a health hazard. Gasoline-driven vehicles are the main source of lead in the environment. A report in 1996 indicated that annual average Pb levels in Mumbai ranged from 0.5 μg m^{-3} to 1.3 μg m^{-3}. In Lucknow, atmospheric Pb levels as high as 2.96 μg m^{-3} were recorded along road sections (N. Singh *et al.*, 1995). In Delhi the Pb concentration in air exceeded the permissible limit of 1.0 μg m^{-3}.

4 Effects of air pollution on forests in India

Forest decline has been a cause of concern in many developed countries. Research has also started on this problem in India, with the aim of determining the causes of mortality of trees and changes in their growth. Air pollutants contribute to forest damage in many parts of the country, especially around industrializing locations. However, there is no information available on the regional impacts of air pollution on forest health.

The research that has been conducted is mainly based on survey-oriented field studies, long-term field studies or artificial exposure studies.

4.1 Survey-oriented field studies

Measurement of spatial and temporal visible injury patterns in plants has been a common practice for this type of study. Plant susceptibility has been identified on the basis of injury symptoms (Pandey, 1978; Bedi *et al.*, 1982; Dubey, 1990).

A case study conducted during 1975 and 1976 in the forest around Obra thermal power plant (550 MW capacity), located in the Sonbhadra district of Uttar Pradesh, revealed leaf injury symptoms in the form of irregular chlorosis and necrosis in the majority of tree species growing near the power plant (Pandey, 1978). Based on the extent of foliar injury and distance of occurrence (i.e. near, far and farthest from the power plant), the forest trees were grouped into very tolerant, moderately tolerant and sensitive categories (Table 9.2). High levels of foliar injury in trees close to an emission source, and gradually reducing injury levels with increasing distance from the source, have also been observed along a transect in Mumbai (Chaphekar, 1972; Chaphekar *et al.*, 1980) and on tree species growing in the Nagda industrial complex in Madhya Pradesh (Pawar and Dubey, 1982, 1983). Flowering and fruiting in mango (*Mangifera indica*) were either reduced or completely absent. Injury from particulate emissions released by a coal gravity trans-shipment yard in Varanasi has been reported on mango, taking the form of defoliation of branches and mortality of apical buds leading to the development of an asymmetric tree canopy (Rao, 1971). Abscission of flowers and reductions in fruit yield have also been noted.

4.2 Long-term field studies

Long-term investigations have been conducted either in forests around a point source of pollution or in plantations of trees growing in urban areas. The results of a long-term study conducted downwind of the 550 MW thermal power plant in Obra indicated changes in the physical and chemical

Table 9.2. Relative sensitivity of some plant species occurring near Obra to the thermal power plant emissions (after Pandey, 1978).

Very tolerant	Moderately tolerant	Sensitive
Acacia catechu Willd.	*Anogeissus latifolia* Wall.	*Aegle marmelos* Correa.
Cassia tora L.	*Boerhaavia diffusa* L.	*Bauhinia tomentosa* L.
Cyperus rotundus Miq.	*Butea frondosa* Wall.	*Boswellia serrata* Roxb.
Dicanthium annulatum (Forskal) Stapf	*Cassia fistula* L.	*Buchanania lanzan* Spreng.
Diospyros melanoxylon Blume	*Desmodium triflorum* DC.	*Gardenia turgida* Roxb.
Eragrostis tenella Nees	*Eclipta alba* Hassk.	*Grewia tiliaefolia* Vahl.
Euphorbia hirta L.	*Hardwickia binata* Roxb.	*Madhuca indica* J.F. Gmel.
Lagerstroemia parviflora Roxb.	*Malvastrum tricuspidatum* A. Gray	*Miliusa tomentosa* (Roxb.) J. Sinclair
Saccharum munja Roxb.	*Phyllanthus simplex* Retz.	*Nyctanthes arbor-tristis* L.
Tephrosia purpurea Pers.	*Scoparia dulcis* L.	*Phaseolus* sp.
Ziziphus jujuba Lam.	*Sida acuta* Burm.	*Phyllanthus emblica* Gaertn.
Ziziphus nummularia DC.	*Vernonia cinerea* Less.	*Setaria glauca* Beauv.

characteristics of the soils (Pandey, 1983). The pH of the soil was 5.0 at 0.5 km NE, 5.8 at 6.0 km NE and 7.2 at 20 km SE of the source. Organic matter accumulation was greater close to the power plant as compared with distant sites. Close to the source, total sulphur and exchangeable potassium increased whereas total nitrogen and available phosphorus declined. Reduction in the chlorophyll and energy (calorific value) contents of leaves and the percentage of photosynthetically active leaf area and an accumulation of sulphur and potassium were observed towards the emission source.

A gradient in species richness was observed at the Obra thermal power plant, with the number of species decreasing towards the source (Pandey, 1978). The numbers of herb, grass, shrub and tree species were 17, 7, 6 and 4 at 3 km NE, 24, 11, 6 and 9 at 9.5 km NE and 24, 15, 6, and 19 at 20 km SW of the source. The reductions in species diversity were attributed to emissions from the thermal power plant.

Reductions in stem circumference and the leaf weight of tree species have also been reported around the Nagda industrial complex (Pawar, 1982) and in the Betul forest area around the Satpura thermal power station (Dubey *et al.*, 1982), located in Madhya Pradesh. Response patterns varied between plant species. *Bridelia retusa* was the most sensitive (Trivedi and Dubey, 1983). Vij *et al.* (1983) have also found that the loss of photosynthetic pigment in the leaves of *Bridelia retusa, Mangifera indica, Tectona grandis, Cassia fistula* and *Adina cardifolia* as induced by air pollution is greater than in *Aegle marmelos.* Dubey (1990) conducted a detailed study in the forested area of south-west Madhya Pradesh and south-east Rajasthan to assess plant responses to air quality. SO_2 and SPM were identified as the most important air pollutants in the region. Various physiological and biochemical parameters were quantified to evaluate response patterns that directly corresponded to pollutant concentrations. The differential responses of plants to pollutants were attributed to the differential adaptive potentials of plants (Rao and Dubey, 1990). *Dalbergia sissoo* enhanced its antioxidative capacity by increasing peroxidase and superoxide dismutase, which in turn may have caused faster sulphite oxidation. A list of plant species, in decreasing order of tolerance, was suggested based on their antioxidant contents.

Pandey (1993) measured biomass accumulation, sulphur pool and inputs into herbaceous vegetation at different distances from the Obra thermal power plant. Peak above-ground biomass was four times greater 10 km downwind as compared with 1.5 km from the source. Sulphur contents were greatest in the soil, followed by the below-ground plant parts and the above-ground plant parts. Sulphate contents were lower in July than October. The study revealed significant reductions in peak standing biomass in response to pollutants.

Pal (1974) studied the forest vegetation around an aluminium smelter at Renukoot in Mirzapur and observed decreases in chlorophyll content and dry weight of leaves of *Shorea robusta*, an important timber tree, in response to fluoride. Later studies involving the assessment of the effects of fluoride emissions on other economically important tree species such as *Buchanania lanzan*, *Terminalia tomentosa* and *Diospyros melanoxylon* in the same area confirmed these findings (Pandey, 1979; Lal and Ambasht, 1981). Peak foliar fluoride levels ranged from 286 ppm in *D. melanoxylon* to 938 ppm in *T. tomentosa*. Rao and Pal (1978) have reported accumulation of fluoride and organic matter in surface soil around the aluminium factory and correlated it with a fluoride-induced decrease in the growth and activity of soil microorganisms. An examination of a large population of cattle feeding on fluoride-contaminated plants around the factory revealed that fluoride-induced symptoms, such as stains, pittings and/or abrasions, had developed to varying degrees on their incisor teeth (Pal, 1974).

A 2-year study conducted in the urban environment of Varanasi to evaluate the responses of a shrub (*Carissa carandas* L.) and two tree species (*Delonix regia* Rafin and *Cassia fistula* L.) to ambient air quality revealed reductions in plant height, basal diameter, canopy area, plant biomass and chlorophyll, ascorbic acid and N contents in leaves at sites receiving higher pollution loads (SO_2, O_3, NO_2 and SPM) (Pandey and Agrawal, 1994).

Several studies conducted under field conditions in different parts of the country to evaluate the effects of cement dust on plants have demonstrated that cement dust deposition to the canopy adversely affects the trees (Sree Rangasamy *et al.*, 1973; Singh and Rao, 1978; Pathmanabham *et al.*, 1979; Lal and Ambasht, 1980; Khanam, 1991). Lal and Ambasht (1980) reported reductions in individual leaf biomass and energy levels and an increase in the concentrations of minerals in leaves coated with cement dust. Jafri *et al.* (1979) studied the effects of cement dust on the epidermal characters of leaves of *Syzygium cumini* and reported a decrease in epidermal cell size, but increases in stomatal and epidermal cell frequency. An increase in trichome density of *Psidium guajava* has also been reported in response to exposure to cement dust (Yunus and Ahmad, 1980). Sree Rangaswamy *et al.* (1973) noted alterations in composition and frequency of plant ecotypes near a cement factory in Coimbatore, Tamil Nadu.

In a study conducted around a petroleum refinery at Barauni, Bihar, *Bambusa bambos* showed lower leaf injury compared with *Bombax ceiba* (Prasad *et al.*, 1979). Irregular, bifacial, interveinal chlorotic and necrotic symptoms were observed on the leaves of mango (*Mangifera indica*) and teak (*Tectona grandis*) located north-east of the refinery (the prevailing wind direction) (Prasad and Rao, 1985). The apical buds of mango were injured, initiating dieback and defoliation. Leaf area injury was directly correlated with dust deposition and foliar sulphur content.

4.3 Artificial exposure studies

Exposure studies have been conducted in chambers under field conditions or growth chambers to examine the relative response of forest tree species to known levels of air pollutants. Farooq and Beg (1982) found a greater increase in cell permeability of sensitive species when treated with aqueous SO_2. Dubey *et al.* (1982) found reductions in biomass accumulation of 1.4–23.7% when *Butea monosperma* tree saplings were exposed to 1.0–3.0 ppm SO_2 (2620–7990 $\mu g\ m^{-3}$), respectively, for 15 days. Farooq *et al.* (1988) conducted experiments with SO_2 exposures from 0.8 $\mu l\ l^{-1}$ (2088 $\mu g\ m^{-3}$) to 19.5 $\mu l\ l^{-1}$ (51,090 $\mu g\ m^{-3}$) for 4 h on several tree species and classified them for their sensitivity on the basis of visible symptoms, SO_2 sorption and various chemical constituents. Beg and Farooq (1988) further suggested an increase in the oxidative state of plants resulting from SO_2 exposure. A detailed exposure study was conducted to evaluate the relative susceptibility of 1-year-old saplings of some trees to 0.3, 0.5 or 1 ppm SO_2 (783, 1305 and 2610 $\mu g\ m^{-3}$, respectively) for 4 h day^{-1}, 5 days a week, for 6 months (Rao *et al.*, 1983). *Azadirachta indica* and *Polyalthia longifolia* had lower levels of change in chlorophyll, ascorbic acid, starch and protein contents than *Diospyros sissoo*, *Euginea jambolana* or *Artocarpus heterophyllus*. Dubey *et al.* (1982) found reductions in dry weight increment and chlorophyll content of *Madhuca indica*, *T. grandis* and *Butea monosperma* sprayed with 2, 4 and 6 g fly ash $m^{-2}\ day^{-1}$. Farooq *et al.* (1985) have shown that exposure of *Haloptelea integrifolia* to acute concentrations of SO_2 caused negative alterations in metabolism, resulting in reduced biomass accumulation and growth.

5 Effects of air pollution on tropical dry deciduous forest: a case study

This section reviews the results of a long-term study conducted to evaluate the response of dry tropical forest to increasing industrialization.

5.1 Study area

The study area is located in the Sonbhadra district of Uttar Pradesh, between 24°15′ and 24°22′N latitude and 80°30′ and 82°40′E longitude, with an altitude ranging from 225 to 450 m above sea level. The climate is tropical monsoonal with three distinct seasons: winter (November–February), summer (April–mid-June) and rainy (late June–Sept). Mean monthly minimum temperature during the study period varied from 6.4 to 28°C and mean monthly maximum temperature from 20 to 42°C. The annual rainfall averaged 1042 mm, of which 85% occurred between late June and September. The soil of the study area consisted of a residual ultisol with colour ranging from reddish brown to yellowish brown. The native vegetation is a typical mixed deciduous forest. The presence of massive coal reserves and availability of water nearby have encouraged open-cast coal mines, high capacity thermal power plants and other industries in the area. The region has become one of the most rapidly growing industrial centres in India (Rao *et al.*, 1990; Singh *et al.*, 1990). Rapid industrialization and increasing human activities have led to serious land degradation, deforestation and undesirable changes in air, water and soil quality.

5.2 Sources and emissions

The present long-term study was conducted in areas affected mainly by emissions from the Renusagar (RTPP) and Shaktinagar thermal power plants (STPP). These have generating capacities of 285 and 2000 MW, respectively. At RTPP, the stack height is 100 m and an electrostatic precipitator (ESP) has been installed since 1989. At STPP, the stack heights are 220 m and all are equipped with ESP facilities. Several study sites were selected to the north-east (i.e. downwind) of STPP along an 18 km transect that included the RTPP and took into account the prevailing wind direction and topography.

Air quality data indicated that the concentrations of air pollutants did not always show a gradient of decreasing levels with increasing distances from the source (Table 9.3). SPM and SO_2 concentrations were highest 6 km from STPP and 0.5 km from RTPP. The major air pollutants were SO_2, NO_2 and particulates in the form of SPM and settled particulates. The mean annual concentrations of SO_2, NO_2 and SPM varied between 18.9 $\mu g\ m^{-3}$ and

Table 9.3. Annual average sulphur dioxide (SO_2), suspended particulate matter (SPM) concentrations and dust fall rate at selected study sites around a thermal power plant in dry tropical forest.

Site	Distance (km) and direction from the emission source	SO_2 ($\mu g\ m^{-3}$)	SPM ($\mu g\ m^{-3}$)	Dust fall rate ($g\ m^{-2}\ day^{-1}$)
Shaktinagar TPP				
1	3.5 NE	70.4	210.1	1.41
2	5.5 NE	156.5	550.2	2.88
Renusagar TPP				
3	2.0 NE	88.6	194.5	1.70
4	8.0 NE	65.1	155.3	1.32

156.2 $\mu g\ m^{-3}$ SO_2, 20.0 to 75.5 $\mu g\ m^{-3}$ NO_2 and 48.3–583.9 $\mu g\ m^{-3}$ SPM at the least polluted and most polluted sites, respectively (Singh *et al.*, 1994).

5.3 Impacts

The average leaf area of plants in the vicinity of the power plants and other heavily polluted sites was smaller than at less polluted sites (Agrawal and Singh, 1991; Agrawal *et al.*, 1993). Total chlorophyll content and specific leaf area were considerably lower at sites receiving higher pollution loads (Table 9.4). The chlorophyll contents of the leaves were a useful diagnostic tool for subtle pollutant effects in the area (Agrawal and Agrawal, 1989; Agrawal *et al.*, 1991). The physiological status of the plants, as reflected by metabolite contents, was affected by a number of factors, including light, temperature and nutrient status. All these factors are considerably changed by the particulate and gaseous emissions. Foliar sulphur analysis indicated that the tree species present at sites experiencing higher annual SO_2 accumulated more sulphate-S than those at sites with low SO_2 levels (Table 9.4) (Agrawal and Singh, 2000). In evergreen plants, the SO_4^{2-}-S content increased gradually from summer through winter, whereas in the deciduous plants there was a greater increase after the flushing of new leaves during summer. The reduction in N content was greater during winter in deciduous plants. The plants growing closer to the thermal power plant accumulated more trace elements (manganese, iron, cadmium, lead and nickel) than those growing at distant sites. In deciduous plants, winter leaf fall reduced the trace element contents in the leaves during the following summer. Foliar SO_4^{2-}-S/organic S ratios increased at stressed locations while remaining almost constant at the control location. This ratio is a useful indicator of the state of forest trees under S pollution stress (Agrawal and Singh, 2000).

Table 9.4. Changes in selected parameters of plants (% of control) growing at different sites around thermal power plants in a dry tropical area.

Site	Parameter	Plants					
		Mangifera indica	*Psidium guajava*	*Cassia siamea*	*Delonix regia*	*Eucalyptus* hybrid	*B. spectabilis*
Shaktinagar TPP							
1 (3.5 km NE from source)	Total chlorophyll	80	72	75	68	83	85
	Ascorbic acid	83	84	85	72	84	89
	Specific leaf area	74	73	74	64	77	82
	Sulphur content	142	118	128	124	734	108
2 (5.5 km NE from source)	Total chlorophyll	73	70	70	66	80	83
	Ascorbic acid	82	80	80	68	81	86
	Specific leaf area	73	75	74	60	75	83
	Sulphur content	155	124	133	140	150	120
Renusagar TPP							
3 (2.0 km NE from source)	Total chlorophyll	85	84	80	76	85	93
	Ascorbic acid	91	87	86	81	84	91
	Specific leaf area	85	79	80	75	84	83
	Sulphur content	152	124	136	138	152	118
4 (8.0 km NE from source)	Total chlorophyll	90	87	85	86	89	96
	Ascorbic acid	94	91	91	83	92	95
	Specific leaf area	95	84	85	82	92	94
	Sulphur content	116	110	104	120	106	106

A significant negative correlation between foliar SO_4^{2-}-S and total chlorophyll was observed for mango (*M. indica*) ($r = -0.884$, $P < 0.05$), which also showed the greatest reductions in other parameters such as ascorbic acid, protein, starch and total N contents (Singh, 1991; Agrawal *et al.*, 1993). These data on the responses of plants to thermal power plants provide clear evidence of significant growth reductions up to a distance of 10 km from the emission source.

J. Singh *et al.* (1995) also reported significant impacts of thermal power plant emissions on the physical and chemical properties of soils. Soil pH was mostly alkaline. Bulk density increased and porosity decreased at sites receiving higher pollution loads. The organic carbon, sulphate sulphur and exchangeable calcium and potassium increased, while total nitrogen and the N mineralization rate decreased. The values of trace elements such as manganese, iron, cadmium, copper, lead and nickel were higher at sites closer to the power plant.

In a detailed study to analyse the impact of the emissions from thermal power plants on the structure of communities, it was found that the number of individuals and species richness increased with decreasing pollution load (Table 9.5). The importance value index (IVI) revealed the dominance of a few species at the sites receiving higher pollution loads. Some plant species were completely absent at heavily polluted sites, but were dominant at the least polluted site. Few plant species, however, were uniformly distributed. The species with higher IVI values at heavily polluted sites were termed resistant and those with high IVI values at distant sites were classified as sensitive. The variations in sensitivity were linked to the different strategies and life forms of the species (Singh *et al.*, 1994). The Shannon–Wiener index of species diversity, species richness and evenness were inversely related, whereas concentration of dominance was directly related to the pollution load in the area (Table 9.5). The changes in species diversity have led to an increase in the proportion of resistant herbs and grasses showing a tendency towards a definite selection strategy within the ecosystem in response to air pollution.

Table 9.5. Species richness, species diversity (H′), concentration of dominance (Cd) and equitability (Es) at selected study sites around thermal power plants in a dry tropical area.

Site	Species richness	Species diversity (H′)	Cd	Es
Shaktinagar TPP				
1 (3.5 km NE from source)	7.66	4.11	0.07	0.93
2 (5.5 km NE from source)	7.25	3.90	0.09	0.88
Renusagar TPP				
3 (2.0 km NE from source)	8.76	4.20	0.07	0.94
4 (8.0 km NE from source)	8.17	4.21	0.06	0.94

The impact of fluoride on the growth of forest trees was also studied around the Hindustan Aluminium Company (Hindalco) situated at Renukoot in the central part of the Sonbhadra district. During the study period, the annual production of Hindalco was 126,000 t^{-1} of aluminium metal and 260,000 t of alumina used in captive production. Pure aluminium is produced by the Hall Heroult process. The original vegetation in the area is a dry deciduous forest type. Gaseous emissions include HF, SO_2 and other gases. Particulate emissions include aluminium fluoride, calcium fluoride and others.

The study conducted around the Renukoot aluminium factory revealed high concentrations of HF and SO_2 (Table 9.6) (Narayan, 1992; Narayan *et al.*, 1994). The annual average HF concentration was 3.4, 2.4, 0.8 and 0.4 μg m^{-3} at distances of 1, 2, 5 and 11 km, respectively, downwind of the source. SPM and SO_2 concentrations gradually decreased with increasing distance from the source. A marked seasonality was observed for ambient concentrations of SPM, SO_2 and HF, all of which were greatest in winter, and for dust fall, which was greatest in summer. The maximum dust fall rate of 7.8 g m^{-2} day^{-1} was recorded 1 km downwind of the factory.

Significantly higher concentrations of fluoride in leaves up to 9 km downwind of the factory were observed. With the exceptions of *Psidium guajava* and *Cassia siamea*, all tree species showed significant negative correlations between a number of physiological parameters and the fluorine flux in air and distance from the factory. Specific leaf area, total chlorophyll content and individual leaf areas were negatively correlated with ambient HF and SO_2 concentrations and with the fluoride levels in foliar tissues. Plants differed in their responses to the emissions. Some important trees such as *Shorea robusta* × *Diospyros melanoxylon* showed higher levels of sensitivity. The reduction in leaf area caused by the emissions from the aluminium smelter suggests that the lower photosynthetic capacity at polluted sites may cause significant energy constraints for the plants.

Narayan *et al.* (1994) observed reductions in the number of woody species in response to aluminium smelter emissions in the Renukoot area (Table 9.7). Tree density and canopy cover gradually increased with decreasing pollution load. Although it is possible that emission could have suppressed the growth of

Table 9.6. Annual average concentrations of particulate and gaseous air pollutants at different sites to the north-east of the Hindalco aluminium smelter.

	Distance from the source (km)			
Pollutant	2	4	6	11
SPM (μg m^{-3})	456.67	384.00	257.67	214.33
HF (μg m^{-3})	3.49	2.41	0.83	0.44
SO_2 (μg m^{-3})	107.6	73.8	45.1	31.3
Dust fall rate (g m^{-2} day^{-1})	3.24	2.11	1.49	0.77

new colonizers, the individuals that were well established flourished because of the lack of interspecific competition. The species diversity of both herbaceous and woody plants was inversely proportional to the pollution load. The woody layer was more affected than the herbaceous layer (Table 9.8). Communities were heterogeneous at distant sites as the ambient conditions there favoured the survival, growth and regeneration of vegetation and also new arrivals. In contrast, at the sites closest to the emission source, the loss of sensitive species created a niche for more opportunistic tolerant species, resulting in them becoming abundant at polluted sites. The dominance diversity curve showed a change from a log-normal distribution to a geometric series distribution, a pattern that normally indicates a harsher environment.

A litter decomposition study has been conducted on *Shorea robusta* litter collected from a heavily polluted site and a control site to examine the decomposition rate and nutrient release pattern (Narayan, 1992). The loss of weight was faster in fluoride-free decomposing litter. Nutrient loss was also significantly higher for fluoride-free litter than for fluoride-contaminated litter. The study clearly depicted the inhibition of the decomposition process that may eventually affect the entire ecosystem through changes in nutrient cycling. These two long-term studies have shown that emissions from thermal

Table 9.7. Species richness, evenness, β-diversity, Shannon–Wiener diversity index and concentration of dominance for the woody layer at different sites around the Hindalco aluminium smelter.

Site	Total number of trees (ha^{-1})	Canopy cover (%)	Species richness	Evenness	β-diversity	Shannon–Wiener index	Concentration of dominance
I (2 km NE)	294	53	2.33	1.84	1.40	1.44	0.247
II (4 km NE)	1063	91	3.74	1.92	1.49	1.92	0.183
III (6 km NE)	1124	102	4.76	2.02	1.52	2.32	0.128
IV (11 km NE)	3296	142	6.12	1.90	1.57	2.51	0.114

Table 9.8. Species richness, evenness, β-diversity, Shannon–Wiener diversity index and concentration of dominance for the herbaceous layer at different sites around the Hindalco aluminium smelter.

Site	Species richness	Evenness	β-diversity	Shannon–Wiener index	Concentration of dominance
I (2 km NE)	5.22	2.23	2.21	2.91	0.075
II (4 km NE)	7.90	2.00	2.23	3.00	0.071
III (6 km NE)	10.24	2.04	2.69	3.23	0.064
IV (11 km NE)	10.37	2.06	2.75	3.27	0.050

power plants and aluminium smelters have not only adversely affected plant performance in forests but also the development and organization of the plant communities in the dry tropical deciduous forest area of India.

In another study, 6-month-old tree saplings of *Mangifera indica*, *Psidium guajava* and *Syzygium cumini* were exposed to 0.08 ppm (208 μg m^{-3}) and 0.15 ppm (390 μg m^{-3}) SO_2 for 2 h day^{-1} for 3 months (Pandey, 1993). The saplings exposed to SO_2 showed reductions in starch, protein, ascorbic acid, total chlorophyll and carotenoid contents and increases in proline and phenol contents and peroxidase activity (Table 9.9). Biomass accumulation was reduced more in the above-ground parts than in the below-ground parts. The greatest alterations in the measured parameters were observed in *M. indica*, which also showed the greatest adverse responses to emissions from thermal power stations.

In an attempt to understand the interactive effects of HF and SO_2 on *Cynodon dactylon*, an important palatable forage species in the forest, plants were exposed to 1.0 μg HF m^{-3}, or 200 μg SO_2 m^{-3}, singly or in combination daily for 2 h from 50 to 170 days in open-top chambers. These exposures had a negative impact on photosynthetic pigments, metabolites and enzyme activity, growth characteristics and biomass accumulation (Narayan, 1992). However, HF and SO_2 in combination counteracted each other. When compared with crops such as *Oryza sativa* and *Vigna radiata*, *C. dactylon* showed very low levels of sensitivity to these pollutants.

6 Conclusions

The air quality data presented in this chapter reveal that air pollution may be having substantial impacts on the vegetation of India. Air pollution levels are

Table 9.9. Percentage increase (+) or decrease (–) in various parameters of tree species exposed to different concentrations of SO_2 for 90 days.

	Mangifera indica		*Psidium guajava*		*Syzygium cuminii*	
Parameters	T_1	T_2	T_1	T_2	T_1	T_2
Chlorophyll	–44	–52	–35	–50	–20	–31
Ascorbic acid	–55	–57	–52	–58	–57	–60
Phenol	+50	+82	+46	+70	+30	+46
Proline	+74	+140	+43	+70	+30	+52
Peroxidase	+120	+170	+95	+135	+46	+85
Average sulphur	+14	+20	+8	+14	+12	+17
Shoot length	–13	–16	–10	–15	–3	–5
No. of leaves	–7	–14	–9	–11	–2	–12
Biomass	–15	–23	–13.5	–21	–7	–13

T_1 = 0.08 ppm for 2 h daily; T_2 = 0.15 ppm for 2 h daily.

rising in some forest areas due to rapid industrialization. Sulphur dioxide and particulates are the most important air pollutants because of increased use of fossil fuels for electricity generation and transport. The meteorological conditions characterizing most of the country are particularly favourable to oxidant formation, but ozone monitoring is very limited.

While a large number of studies have been carried out by various research groups on the responses of forest tree plantations to ambient levels of air pollution, long-term field studies in forested area are very scarce. The number of fumigation experiments undertaken to generate dose–response functions is also very limited. In order to evaluate the responses of various forest types to air pollution, it is important to identify the extent of the problem in areas for which experimental data are not currently available. Small areas at high risk as a result of emissions from point sources and large areas at low risk because of emissions from vehicles and long-range transport of air pollutants need to be identified for different regions of the country.

More detailed national studies are necessary to explore air pollution impacts along with specific issues leading to policy initiatives to reduce the forest loss. The effects of air pollution on forest vegetation could have significant economic, environmental and social impacts as India already has a high rate of deforestation to fulfil the land and other requirements of the growing population. India is a vast country with varying forest types and climate, hence there is a need for a national programme on the influences of air pollution on forests and for the development of a guiding policy for air quality standards in relation to forests.

References

Agarwal, A., Narain, S. and Sen, Srabani (1999) *State of India's Environment: The Citizens Fifth Report. Part I. National Overview*. Centre for Science and Environment, New Delhi.

Agrawal, M. and Agrawal, S.B. (1989) Phytomonitoring of air pollution around a thermal power plant. *Atmospheric Environment* 23(4), 763–769.

Agrawal, M. and Khanam, N. (1997) Spatial and temporal variations in concentrations of particulate matter around a cement factory. *Indian Journal of Environmental Health* 39(2), 97–102.

Agrawal, M. and Singh, J. (1991) Environmental impact around Renusagar Thermal Power Plant. In: Sinha, A.K., Pande, D.P., Kumar, R. and Boojh, R. (eds) *Environment Management in Developing Countries*, Vol. I. *Air and its Management*. The Upper India Publishing House, Lucknow, pp. 147–154.

Agrawal, M. and Singh, J. (2000) Impact of coal power plant emission on the foliar elemental concentrations in plants in a low rainfall tropical region. *Environmental Monitoring and Assessment* 60, 261–282.

Agrawal, M., Singh, S.K., Singh, J. and Rao, D.N. (1991) Biomonitoring of air pollution around urban and industrial sites. *Journal of Environmental Biology* (Special issue), 211–222.

Agrawal, M., Singh, J., Jha, A.K. and Singh, J.S. (1993) Coal and coal-based problems in a dry tropical environment, India. In: Keefer, R.F. and Sajwan, K.S. (eds) *Advances in Trace Elements in Coal and Coal Combustion Residues*. Lewis Publishers, Boca Raton, Florida, pp. 27–57.

Anon. (1987) *The State of Forest Report 1987*. Forest Survey of India, Dehradun.

Anon. (1989) *The State of Forest Report 1989*. Forest Survey of India, Dehradun.

Anon. (1995a) *National Ambient Air Quality Statistics of India, 1992*. Central Pollution Control Board, Delhi.

Anon. (1995b) *The State of Forest Report, 1995*. Forest Survey of India, Dehradun.

Anon. (1997a) *The State of Forest Report, 1997*. Forest Survey of India, Dehradun.

Anon. (1997b) *Ambient Air Quality – Status and Statistics, 1992 to 1995*. Central Pollution Control Board, New Delhi.

Bedi, S.J., Patel, M.D. and Shah, S.J. (1982) Ecomorphological and phytochemical studies to determine the pollutant and its source, causing damage to vegetation. A case study of damage to banana plantation. *Indian Journal of Ecology* 2, 335–338.

Beg, M.U. and Farooq, M. (1988) Sulphur dioxide resistance of Indian trees. II. Experimental evaluation of metabolic profile. *Water, Air, and Soil Pollution* 40, 317–328.

Chaphekar, S.B. (1972) Effect of atmospheric pollutants on plants in Bombay. *Journal of Biological Sciences* 15, 1–6.

Chaphekar, S.B., Boralkar, D.B. and Shetty, R.P. (1980) *Effects of Industrial Pollutants on Plants*. Final Report, U.G.C. Project, New Delhi.

Dubey, P.S. (1990) *Study and Assessment of Plant Response Against Air Pollution in Industrial Environments*. Final technical report (14/260/85 MAB/RE) submitted to Ministry of Environment and Forests, New Delhi.

Dubey, P.S., Trivedi, L., Shringi, S.K. and Wagela, D.K. (1982) *Pollution Studies on Betul Forest Area Due to Satpura Thermal Power Station Aerial Discharges*. Final Report DoEn Project no 19/27/78. Vikram University, Ujjain, India.

Farooq, M. and Beg, M.U. (1982) Effect of aqueous SO_2 on the membrane permeability of common Indian tree leaves. *New Botanist* 7, 213–217.

Farooq, M., Masood, A. and Beg, M.U. (1985) Effect of acute exposure of sulphur dioxide on the metabolism of *Holoptelea integrifolia* plants. *Environmental Pollution* 39, 197–205.

Farooq, M., Saxena, R.P. and Beg, M.U. (1988) Sulphur dioxide resistance of Indian trees. *Water, Air, and Soil Pollution* 40, 307–316.

Jafri, S., Srivastava, K. and Ahmed, K.J. (1979) Environmental pollution and epidermal structure in *Syzygium cuminii* L. Skeel. *Indian Journal of Air Pollution Control* 2, 74–77.

Khanam, N. (1991) Impact of cement factory emission on selected plants and soil. PhD thesis, Banaras Hindu University, India.

Khemani, L.T., Momin, G.A., Rao, P.S.P., Safai, P.D., Singh, G. and Kapoor, R.K. (1989) Spread of acid rain over India. *Atmospheric Environment* 23(4), 757–762.

Khemani, L.T., Momin, G.A., Rao, P.S.P., Vijay Kumar, R. and Safai, P.D. (1995a) Study of surface ozone behaviour at urban and forested sites in India. *Atmospheric Environment* 29(16), 2021–2024.

Khemani, L.T., Tiwari, S., Singh, G., Momin, G.A., Naik, M.K., Rao, P.S.P., Safai, P.D. and Pillai, A.G. (1995b) Acid deposition in the vicinity of a super thermal power plant in India. Terres. *Atmospheric and Ocean Sciences* 6(3), 453–459.

Lal, B. and Ambasht, R.S. (1980) Effects of cement dust pollution on plants of *Psidium guajava*. *Indian Journal of Environmental Health* 22, 231–237.

Lal, B. and Ambasht, R.S. (1981) Impairment of chlorophyll content in the leaves of *Diospyros melanoxylon* in relation to fluoride pollution. *Water, Air, and Soil Pollution* 16, 361–365.

Mahadevan, T.N., Meenakshy, V. and Sathe, A.P. (1984) Some observations on the chemical composition of precipitation in industrial area and its use in air quality assessment. *Mausam* 35, 87–90.

Narayan, Deo (1992) Impact of aluminium factory emission on vegetation and soil. PhD thesis, Banaras Hindu University, India.

Narayan, Deo, Agrawal, M., Pandey, J. and Singh, J. (1994) Changes in vegetation characteristics downwind of an aluminium factory in India. *Annals of Botany* 73, 557–565.

NEERI (1991a) *Air Pollution Aspects of Three Indian Megacities. I: Delhi.* National Environmental Engineering Research Institute, Nagpur, India.

NEERI (1991b) *Air Pollution Aspects of Three Megacities. II: Bombay.* National Environmental Engineering Research Institute, Nagpur, India.

Pal, D. (1974) Effect of fluoride pollution in plants and cattle. PhD thesis, Banaras Hindu University, India.

Pandey, G.P. (1979) Ecophysiological responses of plants to gaseous hydrogen fluoride pollution. PhD thesis, Banaras Hindu University, India.

Pandey, J. and Agrawal, M. (1992) Ozone concentration variabilities in a seasonally dry tropical climate. *Environment International* 18, 515–520.

Pandey, J. and Agrawal, M. (1994) Evaluation of air pollution phytoxicity in a seasonally dry tropical urban environment using three woody perennials. *New Phytologist* 126, 53–61.

Pandey, J., Agrawal, M., Khanam, N., Narayan, Deo and Rao, D.N. (1992) Air pollutant concentrations in Varanasi, India. *Atmospheric Environment* 26B, 91–98.

Pandey, S. (1993) Effect of air pollution on plants in the vicinity of thermal power plant. PhD thesis, Banaras Hindu University, India.

Pandey, S.N. (1978) Effects of coal smoke and sulphur dioxide pollution on plants. PhD thesis, Banaras Hindu University, India.

Pandey, S.N. (1983) Impact of thermal power plant emission on vegetation and the soil. *Water, Air, and Soil Pollution* 19, 87–100.

Pathmanabhan, G., Padmanabhan, C. and Oblisami, G. (1979) Effect of kiln dust pollution on the weed flora. *Indian Journal of Air Pollution Control* 2, 70–73.

Pawar, K. (1982) Pollution studies in Nagda area due to Birla Industrial Complex discharges. PhD thesis, Vikram University, India.

Pawar, K. and Dubey, P.S. (1982) Effects of air pollutants on photosynthetic pigments of *Ipomea fistula* and *Phoenix sylvestris*. In: *All India Seminar on Air Pollution Control*, Indore, pp. 19–21.

Pawar, K. and Dubey, P.S. (1983) Effects of atmospheric pollutants on the morphology and pigment content of *Mangifera indica* L. In: *Proceedings of the VI World Clean Air Congress: Air Quality*, Paris, France, pp. 501–507.

Prasad, B.J. and Rao, D.N. (1985) Phytomonitoring of air pollution in the vicinity of a petroleum refinery. *Environmental Conservation* 12, 351–354.

Prasad, B.J., Rao, D.N., Singh, N. and Pandey, S.N. (1979) A survey of pollugenic effects on vegetation around a petroleum refinery. In: *Proceedings of the Symposium on Environmental Biology*, Academy of Environmental Biology, Muzaffarnagar, India, pp. 211–216.

Puri, G.S., MeharHomji, U.M., Gupta, R.K. and Puri, S. (1983) *Forest Ecology, Phytogeography and Forest Conservation*, Vol. 1. Oxford and I.B.H. Co., New Delhi.

Rao, D.N. (1971) A study of air pollution problem due to coal unloading in Varanasi. In: Euglend, M.M. and Berry, G.T. (eds) *Proceedings of the II International Air Congress*. Academic Press, New York, pp. 273–276.

Rao, D.N. and Pal, D. (1978) Effect of fluoride pollution on the organic matter content of soil. *Plant and Soil* 49, 653–656.

Rao, D.N., Srivastava, R.S., Rai, R.S., Jain, P., Prasad, B.J., Singh, S.K., Agrawal, M. and Nandi, P.K. (1983) *To Study the Relative Susceptibility of Saplings of Some Common Fruit and Avenue Trees to SO_2 Pollution*. Final Technical Report, UGC Project No. F 23–115/79 (SR II).

Rao, D.N., Agrawal, M. and Singh, J. (1990) *Study of Pollution Sink Efficiency, Growth Response and Productivity Patterns of Plants with Respect to Flyash and SO_2*. Final report submitted to Ministry of Environment and Forest, New Delhi, India (MOE/141/266/85).

Rao, M.V. and Dubey, P.S. (1990) Biochemical aspects (antioxidants) for the development of tolerance in plants under low levels of ambient air pollutants. *Environmental Pollution* 64, 55–66.

Shrestha, R.M., Bhattacharya, S.C. and Malla, S. (1996) Energy use and sulphur dioxide emissions in Asia. *Journal of Environmental Management* 46, 359–372.

Singh, A., Sarin, S.M., Shanmugam, D., Sharma, N., Attri, A.K. and Jain, W.K. (1997) Ozone distribution in the urban environment of Delhi during winter months. *Atmospheric Environment* 31, 3421–3427.

Singh, J. (1991) Responses of plants to thermal power plant emission. PhD thesis, Banaras Hindu University, India.

Singh, J. and Agrawal, M. (1992) Metabolic responses of plants to thermal power plant emissions. In: *Proceedings of the 9th World Clean Air Congress, Montreal, Canada*, Vol. 5. Air and Waste Management Association, Pittsburgh, Pennsylvania, IU-22C 07, 1–9.

Singh, J.S., Singh, K.P. and Agrawal, M. (1990) Environmental degradation of the Obra-Renukoot-Singrauli area, India and its impact on natural and derived ecosystems. *The Environmentalist* 11(3), 171–180.

Singh, J., Agrawal, M. and Narayan, Deo (1994) Effect of power plant emissions on plant community structure. *Ecotoxicology* 3, 110–122.

Singh, J., Agrawal, M. and Narayan, Deo (1995) Changes in soil characteristics around coal fired power plants. *Environment International* 21(1), 93–102.

Singh, N., Yunus, M., Srivastava, K., Singh, S.N., Pandey, V., Misra, J. and Ahmad, K.J. (1995) Monitoring of autoexhaust pollution by roadside plants. *Environmental Monitoring and Assessment* 34, 13–26.

Singh, S.N. and Rao, D.N. (1978) Possibilities of using chlorophyll and potassium contents in plants to detect cement dust pollution. *IPHE Journal* 1, 10–13.

Sree Rangaswamy, S.R., Padmanabhan, C., Jambutingam, R. and Gurunathan, M. (1973) Effect of cement dust on plant ecotypes. *Madras Agricultural Journal* 60, 1776–1778.

Trivedi, L. and Dubey, P.S. (1983) Vegetation damage due to Satpura Thermal Power Station at North Betul Forest Division. In: Rao, D.N., Sharma, T.P. and Singh, U.R. (eds) *Air Pollution – Problems and Perspectives*, pp. 343–350.

Varshney, C.K. and Agrawal, M. (1992) Ozone pollution in the urban atmosphere of Delhi. *Atmospheric Environment* 26B, 291–294.

Varshney, C.K., Agrawal, M., Ahmad, K.J., Dubey, P.S. and Raza, S.H. (1997) *Effect of Air Pollution on Indian Crop Plants.* Final report, Overseas Development Agency Project, Imperial College of Science, Technology and Medicine, London.

Vij, B.M., Trivedi, L., Shevade, A. and Dubey, P.S. (1983) Chlorophyll damage in tree species due to air pollution. *Biological Bulletin of India* 3(3), 193–194.

Yunus, M. and Ahmad, K.J. (1980) Effect of air pollution on leaf epidermis of *Psidium guajava* L. *Indian Journal of Air Pollution Control* 3, 62–67.

The Importance of Woodfuels as a Source of Pollution in Developing and Rapidly Industrializing Countries

10

J.L. Innes

Department of Forest Resources Management, University of British Columbia, Vancouver, Canada

Woodfuels are of critical importance as a source of income and as a source of energy in many parts of the world. When managed properly, wood represents a sustainable source of energy that has an important role to play as a substitute for fossil fuels. However, in many developing countries, woodfuels are burnt indoors using inefficient stoves and insufficient ventilation. This results in the accumulation of potentially toxic gases that can result in serious human health problems. The technology exists to centralize wood burning and to reduce the emissions. Efforts need to be made to introduce this technology over wider areas. Such steps would not only reduce emissions, but would increase the efficiency of wood burning, thereby reducing the total demand for woodfuels and as a consequence reducing pressure on forests.

1 Introduction

Woodfuels are associated with many difficulties, including the time taken for the collection of the fuel, deforestation, adverse health effects and energy inefficiency. As a result, the utilization of woodfuels is often seen as a transitional phase before the adoption of energy derived from other sources, particularly fossil fuels. Despite these problems, woodfuels, including both wood and charcoal, represent the most important source of energy in developing countries (Table 10.1). Biomass fuels (biofuels), which include woodfuels, agricultural residues and animal wastes, account for about 30% of the energy consumption

 Air Pollution and the Forests of Developing and Rapidly Industrializing Regions (eds J.L. Innes and A.H. Haron)

Table 10.1. Domestic consumption, in ktoe (thousand tonnes of oil equivalent), of woodfuels in 1996 (from Denman, 1998).

Country/region	Wood	Charcoal	Black liquor
Africa	157,741	7,937	
Latin America	49,529	5,519	2,561
Asia (excluding China)	210,556	6,319	
China	92,654		
Former USSR	11,911	1,971	3
Non-OECD Europe	2,624	42	
Middle East	835	161	
Non-OECD total	525,850	19,981	4,532

in most developing countries, rising to 80% or more in some countries, such as Ethiopia, Mozambique, Tanzania, Zaire, Nepal, Bangladesh and Myanmar (Vergnet, 1986; Bourdaire, 1998). They are the main source of household energy for about 2 billion people (Bourdaire, 1998). More than 80% of the rural population and 20% of the urban population (i.e. 75% of the global population) rely on biofuels to satisfy their cooking energy needs (Davidson, 1998). In 1993, global energy consumption from biomass was 19,134 PJ (10^{15} J) with 47% in Asia, 25% in Africa, 19% in Latin America and 1% in Oceania (Davidson, 1998).

Differences exist in the terms woodfuel and fuelwood. Fuelwood is used to describe wood in the rough that is used to generate energy directly or is used in the production of charcoal. Woodfuels include all fuels derived from wood, including wood in the rough, charcoal and black liquor. However, in international statistics, these different terms are not always distinguished.

Several factors combine to make the use of biomass fuels attractive in developing countries: (i) their accessibility in rural areas where commercial fuels and a centralized electricity grid are unavailable; (ii) restoration of deforested and degraded lands by energy plantations; (iii) saving on foreign exchange spent on oil imports; and (iv) employment generation in energy plantations and rural industries. Of the available biofuels, wood is generally favoured over agricultural residues, and agricultural residues are normally preferred to animal wastes (Marufu *et al.*, 1997). However, this very much depends on the availability and costs of the different fuels.

Global annual carbon emissions from woodfuels are equivalent to 500 million tonnes (Mt) of carbon (Houghton, 1996). In addition, 22 Mt of methane and 0.2 Mt of nitrous oxide are released (Watson *et al.*, 1996). Additional pollutants arising from the burning of woodfuels include aromatics, benzene, benzo[*a*]pyrene, carbon monoxide, formaldehyde and respirable particulate matter. However, there are relatively few global estimates of the emissions of these pollutants, as there are major uncertainties associated with the collection of statistics on biomass use.

When looking at individual countries, the contribution of woodfuels to the greenhouse gas emissions of that country may be substantial. Marufu *et al.* (1999) supplied data for Zimbabwe that illustrate the scale of the problem. National emissions of greenhouse gases emitted from wood burning amounted to 4.6 Tg CO_2-C year^{-1}, 0.4 Tg CO-C year^{-1}, 5.3 Gg NO-N year^{-1}, 14.5 Gg CH_4-C year^{-1}, 24.2 Gg non-methane hydrocarbons (NMHC-C) year^{-1}, 2.9 Gg organic acid-C year^{-1} (acetic and formic acids) and 48.4 Gg aerosol-C year^{-1}. These represent 41% ± 6% of the national CO_2 emissions, 67% ± 6% of the CO emissions and 8% ± 1% of the NO emissions.

Although woodfuels offer a number of environmental benefits over fossil fuels, they can represent a serious health hazard. Many woodfuels are burnt using inefficient stoves and in enclosed spaces. This results in the accumulation of potentially dangerous gases that can adversely affect human health. This chapter provides a brief overview of the problems that may be encountered.

2 Wood versus charcoal

Rural populations in developing countries tend to rely on wood as their primary source of energy. Urban populations tend to use charcoal, as local sources of wood are largely exhausted and charcoal is easier to transport for long distances (Brocard *et al.*, 1996). The volume of fuel is reduced by about 80–85% during conversion to charcoal, and the energy density of charcoal is *c.* 29 MJ kg^{-1}, compared with 7 15 MJ kg^{-1} for wood. This makes the transport of charcoal much more efficient, hence its higher use in urban areas. Charcoal is also increasingly being used for industrial purposes, including the smelting of pig iron (Andreae, 1991). In some cases, the wood for such production comes from conifer plantations; in others, it is assumed that wood will be derived from the surrounding forest. Pig iron production was responsible for the destruction of almost two-thirds of the forests in the state of Minas Gerais, Brazil (Andreae, 1991).

During charcoal production, a number of by-products are produced, including water, carbon monoxide, methanol, tar and volatile products (Table 10.2). Although some of these could be utilized, they are mostly released into the atmosphere (Andreae, 1991). Charcoal is a more efficient fuel for use in stoves, and results in less pollution. Smoke density is higher during wood combustion in a stove than when charcoal is used (Ndiema, 1998).

3 Emissions

The primary concern associated with the use of woodfuels has been its effect on the health of human populations (see below). However, the emissions also

Table 10.2. Emission factors for atmospheric carbon (g C kg^{-1} dry wood) and nitrogen compounds (g N kg^{-1} dry wood) from domestic combustion (from Brocard *et al.*, 1996).

	Charcoal making	Charcoal burning	Firewood
C(CO_2)	120 ± 25	170 ± 40	400 ± 70
C(CO)	30 ± 8	25 ± 10	30 ± 10
C(CH_4)	8 ± 2	0.5 ± 0.4	1.5 ± 0.6
C(NMHC[a])	2.0 ± 0.6	0.10 ± 0.03	2.5 ± 1.3
C(OA[b])	0.2 ± 0.2		0.3 ± 0.2
C(Aer[c])	4.0 ± 1.5		5 ± 3
N(NO_x)	0.02 ± 0.02		0.69 ± 0.18
N(NH_3)	0.07 ± 0.1		
N(N_2O)	0.02 ± 0.04		

[a]Non-methane hydrocarbons (C_2–C_{10}).
[b]Organic acids (formic and acetic).
[c]Aerosols.

have important effects on both global climate and local and regional air quality. Much of what is known was gathered during the CHARCOAL/DECAFE-92 (Lacaux *et al.*, 1994) and FIREWOOD/DECAFE-94 experiments conducted in Côte d'Ivoire (Brocard *et al.*, 1996). These studies generated data on a number of emissions, reproduced in Table 10.2. The emissions differ markedly, depending on which type of fuel is used. Domestic wood burning has emissions similar to that of wood burning in a forest or savannah fire, as the fire involves a flaming process. Charcoal making involves smouldering combustion, whereas charcoal burning involves glowing combustion. For a given amount of wood, charcoal making and burning release less carbon dioxide but more carbon monoxide and methane than burning the wood directly.

4 Health impacts

Much biomass conversion is inefficient, leading to serious pollution and health problems (Smith *et al.*, 1983; Boleij *et al.*, 1989; Davidson, 1998). Particulate matter and carbon monoxide are important, and frequently above the critical limits determined by the World Health Organization. Carbon monoxide is especially important as it is a colourless, odourless gas that destroys haemoglobin in the blood and at high concentrations (0.7% for two hours) it is lethal. As the fuels are primarily used for cooking, women and children tend to be affected more than men (van Horen and Eberhard, 1995; Parikh *et al.*, 1999), although detailed analyses (e.g. Armstrong and Campbell, 1991) reveal that a complex set of interactions (including cigarette smoke and the availability of health care) are present, making simple exposure–response studies difficult. Health problems include acute respiratory infections, pneumonia, tuberculosis, lower birth weights, cataracts and nervous and muscular fatigue (Schwela, 1997). Aldehydes are also released; health effects of aldehydes include mucosal membrane and eye irritation, genetic damage and cancer (Zhang and Smith, 1999).

Differences exist between the impacts of wood burning and charcoal burning. Wood burning produces high volumes of fine particulates, whereas charcoal burning results in greater exposure to carbon monoxide. One study in Zambia indicated that the health of charcoal users and electricity users was not significantly different, whereas the health of wood users, as measured by the peak expiratory flow, was significantly lower than that of either charcoal or electricity users (Ellegard and Egneus, 1992). Other studies have not produced such clear results. For example, an investigation of the indoor pollution associated with wood burning in Malaysia indicated that levels of fine particulates and carbon monoxide were high, but that they were insufficient to be hazardous (Lodhi and Zain-al-Abdin, 1999).

As indicated by the above studies, there is relatively little information about the health impacts of indoor wood smoke, although there is considerably more information available on the human effects of smoke from forest fires. Of the various chemicals that are released during burning, the aldehydes are the most important (Dost, 1991), with formaldehyde and acrolein being particularly important. Emissions from wood burning vary markedly, with Lipari *et al.* (1984) reporting total aldehyde emissions from fireplaces of 600–2300 mg kg^{-1} wood, 20–40% of which were formaldehyde. Formaldehyde causes eye, nose and throat irritation, and is both a carcinogen and mutagen.

Smoke particles are also important. The chemical composition of the particles is extremely complex, as they consist largely of condensed volatile organics which interact with many other materials present in the atmosphere. Consequently, the direct health effects of smoke are extremely difficult to determine (Dost, 1991).

One way to reduce indoor pollution from biomass fuels is to increase the efficiency of the stoves that are used. Such a policy has been introduced in China, resulting in average cooking efficiencies of domestic coal and firewood stoves increasing from 12% in 1979 to 22% in 1995 (Cui Shuhong, 1998). However, the improvements in air quality depend on the design of the stoves and the pollutant being measured. For example, Ballard-Tremeer and Jawurek (1996) compared the emissions of an open fire built on the ground, an improved fire built on a raised grate and three stove designs. The lowest smoke emissions came from the improved open fire, and the lowest emissions of sulphur dioxide and carbon monoxide were from the two open fire designs.

According to Rossier and Micuta (1999), it would be possible to save 90% of the wood used in charcoal making if the wood was burnt directly in fuel-efficient stoves rather than being converted to charcoal and burnt in traditional or improved braziers. In Africa, where over 90% of households use open fires for cooking (Kgathi and Zhou, 1995), the introduction of stoves would significantly reduce indoor pollution problems. However, many programmes to introduce stoves have failed, primarily because insufficient attention is given to issues such as the source and costs of fuels, the problems associated with stove use and the other benefits associated with open fires (e.g. space heating, a measure of insect control and lighting). The most successful programmes have been in areas where woodfuel prices or collection times are high, indicating the importance of assessing the local situation before trying to introduce stoves (Barnes *et al.*, 1993, 1994; Wallmo and Jacobson, 1998). Another approach is to use centralized wood gasification facilities, but such facilities can be relatively costly.

5 Future patterns of biomass consumption in developing countries

The International Energy Agency (IEA) has calculated that the developing countries of the world (defined as non-OECD countries of Africa, Asia and Latin America) currently account for 30% of world consumption of commercial energy. They are expected to be responsible for more than 60% of the increase in energy demand between 1995 and 2020 (Bourdaire, 1998). Globally, biomass energy represents about 15% of the global energy supply (Davidson, 1998), although biofuel consumption may decrease in some countries as they develop and individuals gain access to higher quality sources of energy. This is illustrated by fuelwood consumption in Brazil in comparison with other countries in the region (Table 10.3). In contrast, although the use of kerosene and liquefied petroleum gas (LPG) has increased in India, there has been no reduction in the amount of biofuels used (176 Mt in 1978–1979; Natarajan, 1998), and the decreasing contribution of biomass fuels to the total energy supply actually hides a progressive increase in total fuelwood consumption that is also likely to continue into the future (Natarajan, 1998).

Brazil represents an interesting case, as fuelwood remains an important source of energy. The pattern of its use is changing and the proportion burnt residentially has declined dramatically over the last 20 years. Instead, fuelwood is being used increasingly by industry, using conversion processes that are generally more efficient than those available to domestic users. For example, 41% of fuelwood consumption in Brazil in 1996 was for the generation of charcoal or electricity by the transformation sector. The charcoal is being increasingly sourced from plantations. Currently, fuelwood production in Brazil amounts to 21.9 Mt of oil equivalent (Mtoe) (Melges de Andrade *et al.*, 1998). Biomass in general is a major source of energy, with a further 25.5 Mtoe being supplied by sugarcane.

Table 10.3. Firewood consumption, in kboe (thousand barrels of oil equivalent), in Brazil and other Latin American countries (from Hernandez, 1998).

Year	Brazil	Other Latin American countries
1980	155,560	187,121
1981	154,014	188,303
1982	143,801	192,206
1983	143,794	194,830
1984	149,041	198,172
1985	141,750	201,345
1986	135,421	200,230
1987	137,978	204,042
1988	130,561	206,280
1989	122,292	219,555
1990	111,256	220,559
1991	109,398	225,305
1992	104,241	229,441
1993	98,204	228,349
1994	97,940	216,933
1995	92,823	224,424
1996	92,947	227,459

In many countries, political uncertainty, wars, drought and other problems have created large numbers of refugees. Energy provision in refugee camps is a major problem, especially as refugees often have no money and no source of income. Consequently, attempts to alleviate energy problems through the introduction of fuel-efficient stoves may be unsuccessful, as the refugees are relying on fuel sources such as animal dung (Black and Sessay, 1997).

There is a clear relationship between income and woodfuel use. As incomes rise, use of woodfuels declines as alternative forms of energy become more affordable. This has been illustrated by Bensel and Harriss (1996) in a study of domestic fuel sources in Cebu City, the Philippines. They found that LPG was increasingly substituted for wood as income levels rose. The lowest income households (monthly income < 2000 Philippine pesos) received their energy primarily from woodfuels (65%) and kerosene (16%), with only 5% derived from LPG. Households with monthly incomes of more than 20,000 Philippine pesos did not use wood, and obtained 82% of their energy needs from LPG. If these results are typical, then it seems likely that if incomes rise, the use of woodfuels will be reduced in favour of cleaner forms of energy. However, as with the problems concerning the introduction of improved stoves, a number of social and economic factors may slow down the move away from woodfuels.

6 Conclusions

There appears to be a contradiction between the wishes to substitute wood for fossil fuels (as a means of reducing carbon emissions) in some countries and the wishes to substitute fossil fuels for wood (as a means of reducing deforestation and indoor air pollution) in others. Wood remains the primary source of domestic energy in many developing countries, and the collection of woodfuel is a significant source of income for many people in such countries (e.g. Mgeni, 1996; Venkata Ramana, 1997). A number of steps could be taken to reduce exposure to the emissions from wood burning, including the use of more efficient stoves, better ventilation of cooking areas and conversion to centralized wood gasification.

Experience suggests that the introduction of new technology such as improved stoves is fraught with problems, and for a number of reasons it may not be readily accepted (e.g. Kaur, 1997; Wallmo and Jacobson, 1998). Consequently, a better understanding is required of the factors hindering the adoption of better technology, particularly on the part of aid agencies (cf. Campbell, 1994; Mannan, 1996). Donors need in future to place much more attention on the solution of long-term socio-political processes and problems rather than seeking to provide short-term solutions through inappropriate technology (Campbell, 1994). When new technology is successfully introduced, as with new stoves in Kenya (Schwela, 1997), the improvements in health can be dramatic.

References

Andreae, M.O. (1991) Biomass burning: its history, use, and distribution and its impact on environmental quality and global climate. In: Levine, J.S. (ed.) *Global Biomass Burning. Atmospheric, Climatic and Biospheric Implications.* MIT Press, Cambridge, Massachusetts, pp. 3–21.

Armstrong, J.R.M. and Campbell, H. (1991) Indoor air pollution exposure and lower respiratory infections in young Gambian children. *International Journal of Epidemiology* 20, 424–429.

Ballard-Tremeer, G. and Jawurek, H.H. (1996) Comparison of five rural, wood-burning cooking devices: efficiencies and emissions. *Biomass and Bioenergy* 11, 419–430.

Barnes, D.F., Openshaw, K., Smith, K.R. and van der Plas, R. (1993) The design and diffusion of improved cooking stoves. *World Bank Research Observer* 8(2), 119–141.

Barnes, D.F., Openshaw, K., Smith, K.R. and van der Plas, R. (1994) *What Makes People Cook with Improved Biomass Stoves? A Comparative International Review of Stove Programs.* World Bank Technical Paper No. 242. World Bank, Washington, DC.

Bensel, T.G. and Harriss, R.C. (1996) Biomass fuel utilization in tropical Asia. In: Levine, J.S. (ed.) *Biomass Burning and Global Change.* MIT Press, Cambridge, Massachusetts, pp. 671–681.

Black, R. and Sessay, M.F. (1997) Forced migration, environmental change and woodfuel issues. *Environmental Conservation* 24, 251–260.

Boleij, J.S.M., Ruigewaard, P., Hoek, F., Thairu, H., Wafula, E., Onyango, F. and De Koning, H. (1989) Domestic air pollution from biomass burning in Kenya. *Atmospheric Environment* 23, 1677–1681.

Bourdaire, J.-M. (1998) Biomass energy: the role of the IEA. In: OECD (ed.) *Biomass Energy: Data, Analysis and Trends*. Organization for Economic Cooperation and Development, Paris, pp. 9–11.

Brocard, D., Lacaux, C., Lacaux, J.-P., Kouadio, G. and Yoboué, V. (1996) Emissions from the combustion of biofuels in western Africa. In: Levine, J.S. (ed.) *Biomass Burning and Global Change*, Vol. 1, *Remote Sensing, Modeling and Inventory Development, and Biomass Burning in Africa*. MIT Press, Cambridge, Massachusetts, pp. 350–360.

Campbell, J. (1994) Constraints on sustainable development in Ethiopia: is there a future for improved wood-stoves? *Public Administration and Development* 14, 19–36.

Cui Shuhong (1998) Biomass energy for rural development in China. In: OECD (ed.) *Biomass Energy: Data, Analysis and Trends*. Organization for Economic Cooperation and Development, Paris, pp. 233–240.

Davidson, O.R. (1998) Conclusions of last year's workshop: a starting point. In: OECD (ed.) *Biomass Energy: Data, Analysis and Trends*. Organization for Economic Cooperation and Development, Paris, pp. 13–16.

Denman, J. (1998) IEA biomass energy data: system, methodology and initial results. In: OECD (ed.) *Biomass Energy: Data, Analysis and Trends*. Organization for Economic Cooperation and Development, Paris, pp. 19–37.

Dost, F.N. (1991) Acute toxicology of components of vegetation smoke. *Reviews of Environmental Contamination and Toxicology* 119, 1–46.

Ellegard, A. and Egneus, H. (1992) *Health Effects of Charcoal and Woodfuel Use in Low-Income Households in Lusaka, Zambia*. Energy, Environment and Development Series – Stockholm Environment Institute No. 14, 78 pp.

Hernandez, G. (1998) Biomass energy in Latin America and the Caribbeans: estimates, analysis and prospects. In: OECD (ed.) *Biomass Energy: Data, Analysis and Trends*. Organization for Economic Cooperation and Development, Paris, pp. 67–86.

Horen, C. van and Eberhard, A. (1995) Energy, environment and the rural poor in South Africa. *Development Southern Africa* 12(2), 197–211.

Houghton, R.A. (1996) Converting terrestrial ecosystems from sources to sinks of carbon. *Ambio* 12, 267–272.

Kgathi, D.L. and Zhou, P. (1995) Biofuel use assessments in Africa: implications for greenhouse gas emissions and mitigation strategies. *Environmental Monitoring and Assessment* 38, 253–269.

Lacaux, J.P., Brocard, D., Lacaux, C., Delmas, R., Ahoua, B., Yoboué, V. and Koffi, M. (1994) Traditional charcoal making: an important source of atmospheric pollution in the African tropics. *Atmospheric Research* 35, 71–76.

Lipari, F., Dasch, J.M. and Scruggs, W.F. (1984) Aldehyde emissions from wood-burning fireplaces. *Environmental Science and Technology* 18, 326–330.

Lodhi, M.A.K. and Zain-al-Abdin, A. (1999) Indoor air pollutants produced from fossil fuel and biomass. *Energy Conversion and Management* 40, 243–248.

Mannan, M. (1996) Women targeted and women negated. An aspect of the environmental movement in Bangladesh. *Development in Practice* 6, 113–120.

Marufu, L., Ludwig, J., Andreae, M.O., Meixner, F.X. and Helas, G. (1997) Domestic biomass burning in rural and urban Zimbabwe. Part A. *Biomass and Bioenergy* 12, 53–68.

Marufu, L., Ludwig, J., Andreae, M.O., Lelieveld, J. and Helas, G. (1999) Spatial and temporal variation in domestic biofuel consumption rates and patterns in Zimbabwe: implications for atmospheric trace gas emission. *Biomass and Bioenergy* 16, 311–332.

Maur, D. (1997) Behavioural processes in the adoption of a new technology: case study – transfer of improved cookstoves in village Wazirpur. In: Maheshwari, R.C. and Chaturvedi, P. (eds) *Bioenergy for Rural Energisation.* Concept Publishing Company, New Delhi, pp. 253–258.

Melges de Andrade, A., Morato de Andrade, C.A. and Bodinaud, J.A. (1998) Biomass energy use in Latin America: focus on Brazil. In: OECD (ed.) *Biomass Energy: Data, Analysis and Trends.* Organization for Economic Cooperation and Development, Paris, pp. 87–95.

Mgeni, A.S.M. (1996) Forest products and rural income in Tanzania from a methodological perspective in estimation. *Discovery and Innovation* 8(3), 206–214.

Natarajan, I. (1998) Demand for biofuels in rural households. In: OECD (ed.) *Biomass Energy: Data, Analysis and Trends.* Organization for Economic Cooperation and Development, Paris, pp. 181–191.

Ndiema, C.K.W. (1998) Analysis of wood-stove flue gas. *Science, Technology and Development* 16, 89–103.

Parikh, J., Smith, K. and Laxmi, V. (1999) Indoor air pollution: a reflection on gender bias. *Economic and Political Weekly* 34(9), 539–544.

Rossier, G. and Micuta, W. (1999) Should charcoal braziers be promoted? In: Poker, J. and Stein, I. (eds) *Forests in Focus. Proceedings Forum Forests and Energy, 17–21 January 1998.* Alfred Toepfer Akademie für Naturschutz, Reports, Vol. 12, special issue 1, pp. 62–64.

Schwela, D. (1997) Cooking smoke: a silent killer. *People and the Planet* 6(3), 24–25.

Smith, K.R., Aggarwal, A.L. and Dave, R.M. (1983) Air pollution and rural biomass fuels in developing countries: a pilot village study in India and implications for research and policy. *Atmospheric Environment* 17, 2343–2362.

Venkata Ramana, P. (1997) Rural woodfuel production for urban markets: problems and opportunities in the Cebu province, Philippines. In: Venkata Ramana, P. (ed.) *Rural and Renewable Energy: Perspectives from Developing Countries.* Tata Energy Research Institute, New Delhi, pp. 107–127.

Vergnet, L.F. (1986) *La Place du Bois dans les Preoccupations Énergétique en Zone Inter-Tropicale Africaine.* Séminaire sur la politique énergétique, Lome, Togo, 17–21 février 1986. Centre International de Formation en Politique Énergétique (C.I.F.O.P.E.), Paris.

Wallmo, K. and Jacobson, S.K. (1998) A social and environmental evaluation of fuel-efficient cook-stoves and conservation in Uganda. *Environmental Conservation* 25, 99–108.

Watson, R.T., Zinyowera, M.C. and Moss, R. (eds) (1996) *Climate Change 1995: Impacts, Adaptations and Mitigation of Climate Change. Scientific-Technical Analysis.* Cambridge University Press, Cambridge.

Zhang, J. and Smith, K.R. (1999) Emissions of carbonyl compounds from various cookstoves in China. *Environmental Science and Technology* 33, 2311–2320.

Forest Fires and Atmospheric Pollution

11

R. Atmawidjaja and D.F. Richards

Universitas Pakuan, JL Pakuan, Bogor, Indonesia

Substantial changes are occurring in the ways that forest areas are being used. One major change is the burning of forest lands in an attempt to create additional agricultural lands. The number of human-initiated forest fires has drastically increased in the last few centuries, sometimes creating large forest fires. Large-scale forest fires create long-term terrestrial and atmospheric changes. The greenhouse gases released during burning are not quickly reabsorbed if substantial amounts of forest are removed. After large fires, secondary greenhouse gases such as methane increase in concentration as a consequence of altered microbial activity. Nutrients present in tropical forests are lost to the forest ecosystem during large fires.

Procedures are now available to use satellites to collect global data to detect and determine the pattern of biomass burning. With the use of improved emission models and long-term AVHRR-derived burned-area determinations, better estimates can be made of greenhouse and other trace gases released to the atmosphere. This will give a better understanding of the interaction of forest burning, climate change and geochemical cycles. The total cost of the 1997–1998 Indonesian fires is estimated at US$4.5 billion. As the majority of these fires were set by humans, most of the losses could have been prevented. A list of proactive measures to prevent these losses is presented.

1 Introduction

Throughout most of the earth's 4.5 billion year history, numerous fires have affected the terrestrial land masses. During the last 100–200 years, increasing numbers of fires have been instigated by humans (Levine, 1990, 1991). Many of the world's tropical savannahs are periodically burned. Seventy-five per cent of the African savannahs alone burn annually. This is equivalent to a little more than 300 Mha. The burning of wood for fuel, charcoal production and agricultural waste together with the burning of forests and savannahs consumes about 8680 Mt of dry plant material annually (Seiler and Crutzen, 1980; Crutzen and Andreae, 1990; Hao *et al.*, 1990; Andreae, 1991).

Large-scale biomass burning creates a number of different problems. The burning can cause changes in climate and changes in the chemistry of our atmosphere and can turn forest lands into deserts. The nutrients released by fires are washed out of the soil and lost to the forest ecosystem. All of this changes the way in which the forests and lands are used (Levine, 1996).

2 The detection of biomass burning

The monitoring of forest and grassland fires from space is now both easy and efficient. Several space platforms, such as SPOT, NOAA, Meteosat and Landsat, are available for this function. Most of these platforms use visible and thermal sensing to gather information. These systems provide spectral, spatial and temporal information that permit rapid response to most forest and grassland fires over vast expanses of land mass.

Burn patterns are now documented and placed into historical references using data provided from the Fire in Global Resource and Environmental Monitoring (FIRE) Project which began in 1994. The central focus of FIRE is to document forest and grassland burning for tropical regions and relate the patterns to the use of land. The FIRE Project utilizes the advanced very high resolution radiometer (AVHRR) system on the National Oceanic and Atmospheric Administration (NOAA) satellites for the collection of much of these data. The project also provides input to mobile receiving stations that can be used to verify fires and provide information at local levels as to the size, location and movement of wildfires (Cahoon *et al.*, 1991).

3 Burning of biomass

Biomass is defined as the amount of living material in a unit area or volume, usually expressed as mass or weight. Houghton (1991) estimated that the gaseous and particulate emissions to the atmosphere due to deforestation have increased by three- to sixfold over the last 135 years. In addition, he states that the burning of grassland, savannah and agricultural lands has increased during the last century. Ecosystems such as grassland, savannah and agricultural lands that in the past were rarely burned are now being burned more often. In Latin America, the area of grassland, pasture and agricultural lands increased by about 50% between 1850 and 1985. The same trend of increased burning of ecosystems and an increase in grassland and agricultural areas is true for South and South-east Asia. The trend for an increase in biomass burning is not limited to the tropics, although a number of developed countries now restrict the burning of agricultural wastes because of the impact on air quality.

In analysing 50 years of fire data from the boreal forests of Canada, Russia, other former Soviet countries, the Scandinavian countries and Alaska, Stocks (1991) reported a large increase in the area burned in the 1980s. The largest fire in the recent past destroyed more than 5.1 Mha of boreal forest in China and Russia. This occurred in less than a month during May 1987 (Cahoon *et al.*, 1991). The National Academy of Sciences has given estimates suggesting that global burning is very widespread and may cover 300–700 Mha annually. This corresponds to 2–5% of the land area of our planet.

When we think of human-initiated burning, we usually think only of the burning of tropical rainforest and tropical savannah, with the primary purpose being land clearing. For instance, in Brazil about 8 Mha of tropical rainforests were burned for land clearing in 1987 (Booth, 1989). However, the world's boreal forest comprises a significant portion (at least 25%) of the total forest area. Stocks (1991) and Kasischke *et al.* (1993) reported that an average of 8 Mha of boreal forest were burned annually during the 1980s. Thus, it seems likely that human-initiated fires in boreal forests and other non-tropical areas may contribute meaningful amounts of greenhouse gases to the atmosphere.

4 Burning biomass and its gaseous and particulate emissions

When large amounts of biomass are burned, changes occur in the atmosphere and climate. Burning leads to the formation of greenhouse gases. Those of major concern are CO_2, CH_4 and O_3. The annual average estimates of global emissions from biomass burning are given in Table 11.1.

There are both short- and long-term effects from the burning of biomass on a large scale. When a forest or agricultural biomass burns, all of the CO_2 that has been trapped by these plants over the course of many years is suddenly released and returned to the atmosphere. When forests are destroyed, a very important removal system or sink is also destroyed. When there is little or no regrowth of this vegetation, the CO_2 remains in the atmosphere for many years, with estimates of the atmospheric lifetime of CO_2 ranging from 50 to 200 years (Houghton *et al.*, 1996). When the vegetation is rapidly regrowing, such as in savannah and low brush areas, the CO_2 can be reabsorbed through photosynthesis. The emissions of several other gases that are environmentally

Table 11.1. Estimates of global emissions from biomass burning and global emissions from all sources (Levine, 1990).

Species	Biomass burning (Tg element year^{-1})[a]	All sources (Tg element year^{-1})[a]	Biomass burning (%)
CO_2	3500.00	8700.00	40.00
O_3[b]	420.00	1100.00	38.00
CO	350.00	1100.00	32.00
H_2	19.00	75.00	25.00
NMHC	24.00	100.00	24.00
NH_3	0.51	2.30	22.00
NO_x	8.50	40.00	21.00
NH_3	5.30	44.00	12.00
CH_4	38.00	380.00	10.00
Elemental carbon	19.00	22.00	86.00
Particulate organic carbon	69.00	180.00	39.00
Total particulate matter	140.00	1530.00	7.00

[a]Tg = teragram = 10^{12} g = 10^6 t.
[b]O_3 is not a direct combustion product of biomass burning but results from photochemical reactions involving CO, NMHC and NO_x.

significant but not reabsorbed during photosynthesis can remain in the atmosphere for long periods of time.

Levine (1996) described a method using satellite imagery to estimate the instantaneous emissions of trace gases produced by biomass burning. This method, using the calculation developed by Seiler and Crutzen (1980), is as follows:

> As the total mass $M(C)$ of each carbon species released is related to the mass of the total amount of burned biomass (M), then:
>
> $$M(C) = f \times M$$
>
> where f = the mass fraction of carbon in the biomass to the mass of burned biomass (for CO_2 this is 40% by weight). In order to determine the production of other gases, the emission ratio (ER) has to be determined for each species. The ER for each species is defined as:
>
> $$ER = \frac{\Delta X}{\Delta CO_2}$$
>
> where ΔX is the concentration of species X produced by biomass burning and is equal to $X^* - X$, where X^* is the measured concentration of X in the biomass burn smoke plume and X is the background (out of the plume) atmospheric concentration of the species; ΔCO_2 is the measured concentration in the biomass burn plume, where CO_2 is the background atmospheric concentration of CO_2. This normalizes the concentration of each species in respect to CO_2.

Table 11.2. CO_2 normalized emission ratios for combustion species: summary of field measurements and laboratory studies (in g species kg^{-1} C in CO_2) (Andreae, 1991).

Species	Field measurements	Laboratory studies	'Best guess'
CO	6.5–140	59–105	100
CH_4	6.2–16	11–16	11
NMHCs	6.6–11.0	3.4–6.8	7
Particulate organic carbon (including elemental carbon)	7.9–54		20
Elemental carbon (black soot)	2.2–16		5.4
NO_x	2–8	0.7–1.6	2.1
NH_3	0.9–1.9	0.08–2.5	1.3
N_2O	0.18–2.2	0.01–0.05	0.1
H_2	33		33
SO_x	0.1–0.34		0.3
COS	0.005–0.016		0.01
CH_3Cl	0.023–0.33	0.02–0.3	0.05
O_3	4.8–40		30

As the measurement of CO_2 is quite routine, it leads to relatively easy calculations. Measurements of the emission ratios for CH_4, CO and non-methane hydrocarbons (NMHCs) normalized with respect to CO_2 by using field measurements, laboratory measurements and 'best guess' methods are presented in Table 11.2.

Once the mass of the burned biomass (M) and the species emission ratios (ER) are known, the gaseous and particulate species produced can be calculated. The mass of the burned biomass (M) is related to the area (A) burned in a particular ecosystem by the following relationship (Seiler and Crutzen, 1980):

$$M = A \times B \times \alpha \times \beta$$

where A is the total area burned and B is the average biomass material per unit area in a particular ecosystem, in g m^{-2}; α is the fraction of the average above-ground biomass relative to the total average biomass B, and β is the burning efficiency of the above-ground biomass. The parameters B, α, and β vary with each ecosystem and are determined by assessing the total biomass before and after the burn.

Levine (1996) also discussed a recent discovery that burning results in significantly enhanced biogenic emissions of nitrous oxide (N_2O), nitric oxide (NO) and CH_4. His results emphasize that biomass burning has both a short-term and long-term impact on the production and release of these trace gases to the atmosphere.

If complete combustion of biomass occurred, the carbon would all be converted to CO_2 and water vapour (H_2O) according to the following reaction:

$$CH_2O + O_2 \rightarrow CO_2 + H_2O$$

where CH_2O represents the average composition of biomass. However, complete combustion is never achieved and other carbon compounds such as CO, CH_4, NMHCs and particulate carbon are released. Burning also enhances the biogenic emissions of N_2O, NO and CH_4 as well as the pyrogenic release of various sulphur and bromine compounds. The composition of biomass varies, but it usually contains about 40% carbon, 53.3% oxygen and 6.7% hydrogen. Nitrogen varies from 0.3 to 0.8%; sulphur ranges from 0.1 to 0.9% (Bowen, 1979).

The gaseous emissions change depending on the burning and smouldering phases of the fire. The smouldering phase may last for days, whereas the burning phase may last from minutes to an hour or more. This information is summarized in Table 11.3. The total amount of carbon released is obtained by summing CO_2 + CO + CH_4 + NMHCs + particulate carbon.

5 Microbiological emissions pre- and post-burn

The metabolic activity of microbes in the soils and waters of the world are responsible for a major portion of greenhouse gases (Mooney *et al.*, 1987). Most of these organisms live near the soil–water interface or the soil–air interface. The microorganisms of the earth are the earth's 'cleaning system' and remove many unwanted organic compounds. They are great modifiers of the earth's soils, waters and atmosphere. Soil microorganisms, through their activities in biogeochemical cycling, have major effects on global fluxes of a variety of gases (Dixon *et al.*, 1994).

Some gases are relatively stable while others are reactive. The relatively stable gases that are influenced by microbial activities include CO_2, NO, N_2O, and CH_4. Microorganisms also contribute to the flux of reactive gases such

Table 11.3. Gas production during flaming and smouldering phases of burning based on laboratory experiments (Lobert *et al.*, 1991).

	Percentage in burning stage	
Gas released	Flaming	Smouldering
CO_2	63	37
CO	16	84
NMHCs	33	67
NO_x	66	34
NH_3	15	85
HCN	33	67
CH_3Cl	28	72

as ammonia, hydrogen sulphide and dimethyl sulphide $(CH_3)_2S$, as major examples.

Microorganisms associated with termites digest cellulose. About 0.8% of the cellulose digested by termites is released as CH_4 (Mooney *et al.*, 1987). In tropical wet savannahs and cultivated areas, termite populations are increasing rapidly. This increase is being accelerated by the destruction of tropical forests, which results in the accumulation of dead plant materials on the soil surface. This provides an ideal environment for the growth of termites. Termite activity contributes at least 1.5×10^{14} g of CH_4 per day, together with hydrogen and CO_2, to our atmosphere. A balance once existed between the production of CH_4 by methanogens and its metabolism by methanotrophs. A critical factor influencing CH_4 consumption in soils is the ammonium (NH_4^+) ion concentration (King and Schnell, 1994). As ammonium ion levels increase from burning and other sources, the ability of soils to use CH_4 decreases. When CH_4-oxidizing bacteria are exposed to ammonium, they convert the ammonium into toxic nitrite that further inhibits CH_4 oxidation.

For several years, researchers at the San Dimas Experimental Forest (Los Angeles, USA) have been measuring the concentration of ammonium and nitrate (NO_3^-) in the soil of the chaparral forest, before and after prescribed burns. DeBano *et al.* (1979) and Dunn *et al.* (1979) concluded that burning significantly increased the concentrations of ammonium and decreased the concentrations of nitrate in the soil. DeBano *et al.* (1979) found that immediately after an intense fire, soil associated with chamise (*Adenostoma fasciculatum*) vegetation increased in ammonium concentration from 3.79 to 10.77 kg ha^{-1} and the nitrate concentration decreased from 0.58 to 0.47 kg ha^{-1}.

6 Pre- and post-burn values for methane in forests and grasslands

There is usually an increase in the metabolic activity of methanogenic bacteria following a fire. Some areas show an increase of 100% above the pre-fire production level. There is a 2-day period following the fire during which most of the CH_4 production stops. This is most likely because of the destruction of a large portion of the population of methanogens in the area and the purge of CH_4 by the heat. There is then a rise in production of CH_4 that peaks 3–5 days after the fire. Most methanogens are restricted in their ability to utilize nutrients, thus only a few compounds can be utilized to produce CH_4. A few of these nutrient compounds, such as CO_2, acetate, formate, methanol and methylated amines, are produced by the fire (Cicerone and Oremland, 1988). CO_2, acetate and formate are produced by fires in relatively large amounts, CO_2 in the greatest quantities. Particulates collected by airborne filters in the smoke of fires contain fairly large concentrations of both acetate and formate (Cicerone and

Oremland, 1988). Carbon dioxide is very soluble in water, as are formate and acetate compounds. After a fire, when the smoke particulates fall to earth and CO_2 is reabsorbed, it appears very likely that the stage is set for the increased growth of methanogens and the resultant methogenesis. This increase is rather short-lived, however, because the nutrients are readily used up by the microbes.

7 Burning of a boreal (northern) forest

The boreal forest comprises about 25% of the world's forests. Because of limited coverage by observation satellites, it has been difficult in the past to determine the area burned in these forests. Recently the coverage by satellites has become nearly complete. This coverage indicates that the previously determined average of 1.5 Mha burned annually during the 1980s is actually much closer to 5 Mha on average, with great year-to-year fluctuations.

In the Helongjing Province of China, three human-initiated fires started about 6 May 1987. These fires coalesced and continued over a period of 3.5 weeks and destroyed 1.3 Mha of China's boreal forest (Cahoon *et al.*, 1991). At about the same time the fire crossed over the Amur River into Russia, where approximately 4.1 Mha were burned. That year (1987) a total of 14.4 Mha of forest in China and Russia were burned. This fire is probably the largest recorded in modern history.

8 Burning of the tropical savannah

Tropical savannahs are arid or semiarid grasslands with a sparse cover of trees and shrubs. Savannahs are extensive in the tropical regions. They occupy 65% or 20 million km^2 of the land situated between the moist, equatorial rainforests and the deserts. They have been described as marginal environments, where even minimal human intervention can sometimes cause dramatic impacts.

Global estimates of biomass burning comprise the amount of burned forests, savannahs, crop-land wastes and wood for cooking, heating and charcoal production. The burning of savannahs alone destroys three times more dry biomass annually than the burning of tropical forests (Andreae, 1991). This information is summarized in Table 11.4.

9 Burning of the tropical rainforests

Tropical rainforests occur in areas where there is abundant rainfall throughout the year. The largest areas of tropical rainforest are located in the Amazon Basin in South America, in southern Asia including Indonesia, and in the Congo Basin in Africa. Tropical rainforests contain much of the planet's

Table 11.4. Global estimates of annual amount of biomass burning and the resulting release of carbon and CO_2 to the atmosphere (Seiler and Crutzen, 1980; Crutzen and Andreae, 1990; Hao *et al*., 1990; Andreae, 1991).

Source of burning	Biomass burned (Tg C year^{-1})	Carbon released[a] (Tg C year^{-1})	CO_2 released[b] (Tg C year^{-1})
Savannah	3690	1660	1494
Agricultural waste	2020	910	819
Fuelwood	1430	640	576
Tropical forests	1260	570	513
Temperate and boreal forests	280	130	117
Charcoal	21	30	27
World total	8700	3940	3546

[a]Based on a carbon content of 45% in the biomass material. For charcoal, the rate of burning has been multiplied by 1.4.
[c]Assuming that 90% of the carbon released is in the form of CO_2.

biodiversity, including millions of insect species and many plant species that have not yet been described formally. As rainforests contain the earth's greatest diversity of plants and animals, this huge resource may represent giant gene pools that have the potential to provide new drugs, foods and many other products. A few medicinal substances that have already been discovered in rainforests include diosgenin (an active agent in contraceptive pills), reserpine (used to treat heart problems) and curare (used during heart and lung surgery). Only a very small percentage of rainforest plants have been tested for the presence of valuable chemicals.

The destruction of these forests is much more significant in terms of pollution because it creates two separate problems. Firstly, when the rainforest burns, stored carbon is suddenly released in the form of CO_2, CO, CH_4 and other gases. Because the forests are slow to recover, these pollutants are widely dispersed into the atmosphere. The second problem is that, in addition to the release of polluting gases and the destruction of a valuable carbon sink, the nutrient cycling capability of the forest and the nutrients themselves are lost from the ecosystem.

The soil in a tropical rainforest is for the most part very poor in nutrients. The forest depends on precipitation, weathering and the decomposition of forest litter to provide a source of nutrients for forest growth. The decomposition of forest litter is of great importance. About 90% of the nutrients are usually recycled. When a fire destroys the forest litter and trees, most of the nutrients literally go up in smoke. While some nutrients are returned in the form of particulates and ash, they are very short-lived and can be totally depleted within 1–2 years. The ground where a forest once stood becomes useless unless expensive chemical fertilizers are used to supply the nutrients.

When the forests are gone, the nutrient and hydrological cycles are disrupted. The hydrological cycle is changed, and there is a high probability of an irreversible decline in the available water in the region. In areas where large rainforests are located, the forest recycles a large portion of the available water by transpiration. When the forests are removed, that water will run off and remove most of the soluble nutrients.

In Indonesia many traditional societies have practised slash-and-burn cultivation for many years. These traditional methods have remained successful because most of the areas being burned were relatively small. After a few years, when the soil was depleted, the area was abandoned and then was quickly reclaimed by the surrounding forest.

Since the 1970s, the speed of tropical deforestation has greatly accelerated. Australia and the South-east Asia region, including Indonesia, Malaysia, Melanesia and the Philippines, have begun to suffer huge losses of tropical rainforests as a result of timber and agricultural uses. In Amazonia, huge tracts of land have been cleared for cattle ranches and government programmes for colonization. The total rates of deforestation are unknown but estimates are more than 100,000 km^2 $year^{-1}$. Some unique rainforests have probably been destroyed and will likely never return.

There is growing concern about the loss of biodiversity and global warming caused by increased levels of CO_2 that result from rainforest destruction. In addition to this problem, concern is increasing over the destruction of the ozone layer by methyl bromine. This gas is released when the forest burns; it is ten times more active in the destruction of ozone than chlorofluorocarbons, and 40 times more destructive than chlorine (Mann and Andreae, 1994).

10 Historical and recent Indonesian forest fires

Throughout history, Indonesians have used fire as an inexpensive method to clear land. The normal burning cycle starts soon after the dry season begins. The drought that usually accompanies an El Niño event escalates the dangers associated with fires set to clear land and the fires often get out of control. Wild forest fires have plagued the islands of the Indonesian archipelago since prehistoric times. Evidence of these early wildfires on the island of Kalimantan has been found by Goldammer and Siebert (1990). The evidence they collected indicated that major wildfires had occurred between 17,510 and 350 years before the present.

There are several historical accounts noted in Dutch and Portuguese sailors' and explorers' logbooks that tell of large areas of burning peat swamps sending thick smoke to areas as far away as the southern tip of Malaysia. According to continuous observations during the past 10 years, there seems to be a correlation between these fire events and El Niño – Southern Oscillation

(ENSO) years (Brookfield *et al.*, 1995). In recent times there has been a dramatic increase in the numbers of human-induced fires.

10.1 The Indonesian fires of 1982–1983

The 1982–1983 fires in East Kalimantan were severe. Starting in July 1982 and lasting until April 1983, an exceptionally intense drought occurred in connection with an ENSO event. Vegetation was not yet completely dry when the fires were started in November and December of 1982. The fires burned slowly and were mainly restricted to the forest floor. This drought lasted until rain came in May 1983. The fires were most intense from January 1983 to the end of April 1983. Little could be done to control the fires. There were critical shortages of experienced fire-fighters and very little equipment available. Making things worse, most of the local people believed the fires were beneficial in clearing land for future farming efforts. The fires in the area started from small agricultural fires that escaped into dry secondary and logged forests. Estimates of the extent of the fire were about 3.6 Mha of totally or partially burned land. The total loss in economic value of timber in 1982–1983 was set at US$7.891 billion for logged-over forests and US$0.348 billion for swamp forests, with a total of US$8.239 billion for timber losses (Schindele *et al.*, 1989).

10.2 The Indonesian fires of 1994

During 1994 in Kalimantan and Sumatra, a total of 4.86 Mha burned (Tsuruta, 1997). Smoke and haze caused severe health problems in Singapore, Peninsular Malaysia and elsewhere and brought air traffic at some airports to a halt. Only a comparatively small number of these fires were uncontrolled. The total loss for these fires was estimated at US$15.5 million by the Ministry of Forestry.

10.3 The Indonesian fires of 1997–1998

At the beginning of February 1997, climatologists forecast that an intense and prolonged El Niño event was about to occur. In June and July, the State Environmental and Welfare Ministers both issued warnings of the impending drought. It seemed that few if any people were listening and by September 1997 many large and uncontrolled fires were raging in East Kalimantan and Sumatra. Within a short time thick smoke and haze covered much of East Kalimantan, Sumatra, Malaysia and Singapore.

An estimate from the Ministry of Forestry listed the size of the 1997 fire as 500,000 ha burned. The European Union Fire Response Group (EUFREG) estimates (based on the interpretation of satellite imagery) that 2.3 Mha burned in south Sumatra alone (Ramon and Wall, 1998). The Centre for Remote Imaging, Sensing and Processing (CRISP) estimates that 1.5 Mha burned in Sumatra and 3 Mha burned in Kalimantan (Liew *et al.*, 1998).

With graphic results available from satellite imagery, 133 plantation companies, 27 timber estates and 15 transmigration programmes were given a 15-day ultimatum to stop burning or face the revocation of their licences (Director General of PHPA). PL London Sumatra, a palm oil company, was convicted in March 1998 for setting fires. It is also believed that criminal use of fire may have been a major cause.

Indonesia and other ASEAN countries have suffered staggering losses from the fires as well as from the smoke and haze associated with the fires (see Tables 11.5–11.7). The cost of these fires in timber alone at the end of April had reached US$1.06 billion.

Table 11.5. Fire and haze-related damage from the 1997 Indonesian forest fires (in US$ millions).

Type of loss	Lost to Indonesia	Lost to other countries	Total
Timber	493.7		
Agriculture[a]	470.4		
Direct forest benefits	705.0		
Indirect forest benefits[b]	1077.1		
Capturable biodiversity[c]	30.0		
Fire-fighting costs	11.7	13.4	25.1
Carbon release	–	272.1	272.1
Total fire	2787.9	285.5	3073.4

[a]Losses to plantations and small holdings.
[b]Storm protection, water supply and regulation, erosion control, soil formation, nutrient recycling and waste treatment.
[c]Potential income lost to Indonesia from international conservation expenditures, i.e. from international agencies and NGOs, willing to pay to conserve tropical forests. It does not reflect the cost in loss of biodiversity.

Table 11.6. Summary of haze-related damage from the 1997 Indonesian forest fires (in US$ millions).

Type of loss	Lost to Indonesia	Lost to other countries	Total
Short-term health	924.0	16.8	940.8
Tourism	70.4	185.8	256.2
Other	17.6	181.5	199.1
Total haze	1012.0	384.1	1396.1

Table 11.7. Total cost of the 1997 Indonesian forest fires.

To Indonesia	US$3.8 billion (85%)
To other countries	US$0.7 billion (15%)
Total	US$4.5 billion

The government goal was to clear 0.69 Mha each year for palm oil plantations. Peat forests were also being cleared in 1997 for a huge government project to convert 1 Mha of peat forest into a mega-rice paddy for people moving in the transmigration programme. Most of the forest fires that burned in Indonesia in 1997–1998 were caused by humans. That being the case, they should have been preventable.

What measures can be taken to prevent some of the losses that the fires have brought? The government ministers that are responsible for fire management and fire-fighting policy implementation in Indonesia need to become proactive. With new technology and help from the many sources offering assistance, Indonesia should formulate a fire and haze prevention plan and consider among other items the following:

(A) Fire and haze prevention

1. Data management, the rapid and complete dissemination of information dealing with fire issues such as:

- the types of health hazards associated with the burning area;
- the presence of haze and its movement;
- the type of activities that are going on in the area; and
- the location and number of fires that are burning.

2. Accurate mapping of areas subject to increased risk of fires. This should include the seasonal variation of expansion and contraction due to climatological conditions.

3. The forecasting of climatological conditions that may result in fire and haze events and the movement of the fire and the haze.

4. A review of current policy at the State and lower levels which will determine ways to compensate landowners properly and fairly for their timber and land so that they are not as likely to use fire as a negative tool or a weapon.

5. Development of a market-based or some other economic incentive that promotes the adoption of new products and technologies that can use logging waste, biomass and land-clearing residues for useful materials.

6. The development and implementation of governmental arrangements for linking State and local fire-fighting capability in any combination with ASEAN.

7. Arrangements for the employment of local fire-fighters who can be well trained and are willing to fight fires locally or nationally.

8. Inventory of existing fire-fighting capabilities at the State and local levels, including all aspects of fire-fighting equipment and personnel.

9. Ensuring the continued readiness of fire-fighters' capabilities at the State and local level by means of regular maintenance of equipment and upgrading of personnel skills.

(B) Surveillance of the prevention and reduction of forest fires and haze programmes
1. The early detection and location of forest fires by the latest technology and rapid communication to the local or front-line fire-fighters.
2. Use of new technology to track and predict the size and movement of fires and the resulting haze.
3. The forecasting of the degree to which a fire is likely to generate haze as well as the types of emissions that are likely to be emitted.
4. The determination of which health impacts are likely as a result of a particular haze episode.
5. Keeping historical records of each fire event, how it proceeded and what type of haze was produced.
6. Track the socio-economic cost of each particular fire event, the extent of the area burned and the composition of the flora and fauna destroyed.

In order to prevent future forest fires and their accompanying devastation, there must be a good understanding of all of the overt and underlying causes of these fires. There is only so much blame that can be placed on El Niño for the fires caused by humans. Indonesia must begin to form a clear and strict set of policies that can and will be uniformly enforced. The consequences of any infringement must be so severe that people will not set fire to the public lands for any reason.

11 Conclusions

- The amount of biomass being burned is increasing both from natural effects and from human-induced fires.
- The nutrients released from large fires are not immediately recycled into the forest ecosystems. The gases released contribute to global warming, due to the increase in greenhouse gases.
- Forest burning disrupts normal nutrient and water cycling and may convert lush forest areas into unproductive arid deserts.
- Biogenic gases resulting from fires also contribute to global warming.
- There is increasing pressure to convert rainforests and savannahs into agricultural lands.
- The increased conversion of forests into agricultural lands decreases the high biodiversity normally found in forests and may eliminate species that could provide tremendously important gene pools and extremely useful drugs.

- Improved satellite observation of forest fires is enabling world scientists to better assess the terrestrial and atmospheric changes that are occurring due to forest fires.
- The 1997 Indonesian fires were accentuated by severe drought associated with an El Niño event and land-clearing activities.
- The majority of the 1997 Indonesian fires were set to clear land and as such could have been prevented.
- Proactive measures by Indonesian governmental ministers responsible for fire management could reduce most of the losses caused by fire and haze, by the implementation of strict polices that are rigorously enforced.

Acknowledgements

We appreciate the work of all of the scientists whose contributions are referenced in this review. They deserve the credit and heartfelt thanks of the authors. We also thank Dr Sheril D. Burton for his help with technical assistance and his invaluable contributions to the preparation of this review.

References

Andreae, M.O. (1991) Biomass burning. Its history, use, and distribution and its impact on environmental quality and global climate. In: Levine, J.S. (ed.) *Global Biomass Burning: Atmospheric, Climatic, and Biospheric Implications.* MIT Press, Cambridge, Massachusetts, pp. 3–21.

Booth, W. (1989) Monitoring the fate of the forests from space. *Science* 243, 1421–1429.

Bowen, H.L.M. (1979) *Environmental Chemistry of the Elements.* Academic Press, London.

Brookfield, H.C., Potter, L. and Byron, Y. (1995) *In Place of the Forest: Environmental and Socio-Economic Transformation in Borneo and Eastern Malay Peninsula.* United Nations University Press, Tokyo.

Cahoon, D.R. Jr, Levine, J.S., Cofer, W.R. III, Miller, J.E., Tennille, G.M., Yip, T.W. and Stocks, B.J. (1991) The great Chinese fire of 1987: a view from space. In: Levine, J.S. (ed.) *Global Biomass Burning: Atmospheric, Climatic, and Biospheric Implications.* MIT Press, Cambridge, Massachusetts, pp. 61–66.

Cicerone, R.J. and Oremland, R.S. (1988) Biogeochemical aspects of atmospheric methane. *Global Biogeochemical Cycles* 2, 299–327.

Crutzen, P.J. and Andreae, M.O. (1990) Biomass burning in the tropics: impact on atmospheric chemistry and biogeochemical cycles. *Science* 250, 1669–1678.

DeBano, L.F., Eberlein, G.F. and Dunn, P.H. (1979) Effects of burning on chaparral soils. I. Soil nitrogen. *Soil Science Society of America Journal* 43, 504–509.

Dixon, R.K., Brown S., Houghton, R.A., Solomon, A.M., Trexler, M.C. and Wisniewski, J. (1994) Carbon pools and flux of global forest ecosystems. *Science* 263, 285–190.

Dunn, P.H., DeBano, L.F. and Eberlein, G.E. (1979) Effects of burning on chaparral soils: II. Soil microbes and nitrogen mineralization. *Soil Science Society of America Journal* 43, 509–514.

Goldammer, J.G. and Siebert, B. (1990) The impact of droughts and forest fires on tropical lowland rain forest of East Kalimantan. In: Goldammer, J.G. (ed.) *Fire in the Tropical Biota: Ecosystem Processes and Global Changes*. Springer-Verlag, Berlin, pp. 11–31.

Hao, W.M., Liu, M.H. and Crutzen, P.J. (1990) Estimates of annual and regional release of CO_2 and other trace gases to the atmosphere from fires in the tropics, based on FAO statistics for the period 1975–1980. In: Goldammer, J.G. (ed.) *Fire in the Tropical Biota: Ecosystem Processes and Global Changes*. Springer-Verlag, Berlin, pp. 440–462.

Houghton, R.A. (1991) Up in flames: Kuwait's burning oil wells are a sad test of theories. *Scientific American* 265, 17–24.

Houghton, J.T., Meira Filho, L.G., Callander, B.A., Harris, N., Kattenberg, A. and Maskell, K. (eds) (1996) *Climate Change 1995. The Science of Climate Change*. Cambridge University Press, Cambridge.

Kasischke, E.S., French, N.H.F., Harrell, P., Christensen, N.L. Jr, Ustin, S.L. and Barry, D. (1993) Monitoring of wildfires in boreal forest using large area AVHRR NDVI composite image data. *Remote Sensing of the Environment* 44, 1–10.

King, G.M. and Schnell, S. (1994) Effect of increasing atmospheric methane concentration on ammonium inhibition of soil methane consumption. *Nature* 370, 282–284.

Levine, J.S. (1990) Global biomass burning: atmospheric, climatic and biospheric implications. *EOS, Transactions, American Geophysical Union* 71, 1075–1077.

Levine, J.S. (1991) *Global Biomass Burning: Atmospheric, Climatic, and Biospheric Implications*. MIT Press, Cambridge, Massachusetts.

Levine, J.S. (1996) *Biomass Burning and the Production of Greenhouse Gases*. NASA Langley Research Center Publication.

Liew, S.C., Lim, O.K., Kwoh, L.K. and Lim, H. (1998) A Study of the 1997 Forest Fires in South East Asia Using SPOT Quicklook Mosiacs. Paper presented at the 1998 IEEE International Geoscience and Remote Sensing Symposium, 6–10 July 1998, Seattle.

Lobert, J.M., Scharffe, D.H., Hao, W.M., Kuhlbusch, T.A., Seuwen, R., Warnesk, P. and Crutzen, P.J. (1991) Experimental evaluation of biomass burning emissions: nitrogen and carbon containing compounds. In: Levine, J.S. (ed.) *Global Biomass Burning: Atmospheric, Climatic, and Biospheric Implications*. MIT Press, Cambridge, Massachusetts, pp. 289–304.

Mann, S. and Andreae, M.O. (1994) Emission of methyl bromide from biomass burning. *Science* 263, 1255–1257.

Mooney, H.A., Vitousek, P.M. and Matson, P.A. (1987) Exchange of materials between terrestrial ecosystems and the atmosphere. *Science* 238, 926–932.

Ramon, J. and Wall, D. (1998) *Fire and Smoke Occurrence in Relation to Vegetation and Land Use in South Sumatra Province, Indonesia with Particular Reference to 1997*. Report No. 47, Forest Fire Prevention and Control Project, European Commission.

Schindele, W., Thomas, W. and Panzer, K.F. (1989) *The Kalimantan Forest Fires of 1982–83 in East Kalimantan Part 1: The Fire, the Effects, the Damage and Technical*

Solutions. FR Report No. 5, German Agency for Technical Cooperation and The International Tropical Timber Organization, Jakarta, Indonesia.

Seiler, W. and Crutzen, P.J. (1980) Estimates of gross and net fluxes of carbon between the biosphere and the atmosphere from biomass burning. *Climatic Change* 2, 207–247.

Stocks, B.J. (1991) The extent and impact of forest fires in northern circumpolar countries. In: Levine, J.S. (ed.) *Global Biomass Burning: Atmospheric, Climatic, and Biospheric Implications*. MIT Press, Cambridge, Massachusetts, pp. 197–202.

Tsuruta, H. (1997) Regional air pollution of aerosols and ozone by forest fires in tropical East Asia: a case study of the 1994 episode. In: Haron, A.H., Dahlan, T., Mohd. Puat, D. and Amran, M. (eds) (1997) *Transboundary Pollution and the Sustainability of Tropical Forests*. ASEAN Institute of Forest Management, Kuala Lumpur, pp. 57–61.

Development of the Regional Policy Process for Air Pollution in Asia, Africa and Latin America

12

W.K. Hicks,[1] J.C.I. Kuylenstierna,[1] V. Mathur,[1] V. Burijson,[2] S. Mazzucchelli,[2] M. Iyngararasan,[3] S. Shrestha,[3] and A.M. van Tienhoven[4]

[1]*Stockholm Environment Institute at York, University of York, York, UK;* [2]*Instituto Internacional de Medio Ambiente y Desarrollo (IIED) – America Latina, Buenos Aires, Argentina;* [3]*UNEP Environment Assessment Programme for Asia and Pacific, AIT, Pathumthani, Thailand;* [4]*CSIR Environmentek, Pretoria, South Africa*

Air pollution is an increasingly important environmental problem in developing country regions, particularly in parts of Asia. Projections indicate that potentially large increases in emissions may occur during the next 20–50 years if current development patterns persist. This chapter focuses on the regions of East, South-east and South Asia, southern Africa and the MERCOSUR countries of Latin America. The status and development of regional policy initiatives on air pollution vary from region to region. Regional/sub-regional activities have been developed through different intergovernmental organizations building (in some cases) on the experiences of intergovernmental cooperation in Europe. The initiation of international cooperation in developing country regions can be triggered by dramatic impacts at the regional level, or by increasing awareness in the policy-making community of an escalating problem. Increased contact between international and regional experts and policy makers is considered to be an important factor for developing the motivation to address regional air pollution problems.

1 Introduction

Air pollution is becoming an increasingly important issue in many developing countries. This chapter describes the current state of policy development at the regional scale for air pollution in developing countries and provides examples of how international cooperation is being promoted. 'Regional air pollution' refers to that part of the issue that is not solely local (e.g. in urban areas) or global (e.g. global climate change). In that emissions often give rise to local as well as regional impacts (after long-range transport), local impacts are here considered part of the regional air pollution problem.

Major sources of air pollution include the combustion of fossil fuels in the electricity generation, transport and industry sectors. Other sources include industrial processes, agricultural and other land use practices (such as forest fires and waste handling and disposal). Chemical reactions in the atmosphere involving air pollutants also produce a number of important secondary air pollutants, such as ozone and small particulates, that are responsible for photochemical smog and haze. This chapter concentrates on measures taken to control the 'classic pollutants': the oxides of sulphur and nitrogen, particulates and ozone. Other pollutants, such as the greenhouse gases (GHGs), ozone-depleting substances (ODS) and persistent organic pollutants (POPs), have also been the subject of international protocols but are not discussed here.

In Europe, air pollution has a long history. Initially, problems occurred in major cities, affecting health. In the 1950s, for example, smog disasters in London caused several thousand excess deaths and led to several policy measures outlined in the Clean Air Act of 1956. Smokeless zones were introduced and industry was moved away from urban centres and required to install taller stacks. The development of tall stacks, coupled with the dramatic rate of industrialization and fossil fuel use in Europe, caused the development of regional problems such as ecosystem acidification. Acidification, mainly as a consequence of sulphur emissions, caused a great degree of damage to lake and stream ecosystems in Scandinavia, the UK and central Europe where there was a combination of high sulphur deposition and high ecosystem sensitivity (Battarbee, 1990). Similarly, impacts on lakes were experienced in eastern North America (Elder and Brydges, 1983). In the 1960s it was realized that the problems were caused by transboundary transfer of air pollution, meaning that individual countries could not solve their acidification problems alone. This then necessitated the development of international agreements for coordinated emission control. Eventually, the international policy negotiations of the United Nations Economic Commission for Europe (UN-ECE) led to the Convention on Long-range Transboundary Air Pollution (LRTAP) which has resulted in Protocols for the control of sulphur, nitrogen and volatile organic compound (VOC) emissions (UN-ECE, 1996). The latest UN-ECE Protocol is the 'multi-pollutant, multi-effect' Protocol, signed in Gothenburg, Sweden, in December 1999. In addition, various EU Directives and strategies regarding large combustion plants and acidification have been developed (Bull and Fenech, 1999). The existence of successive stages of awareness and action to deal with air pollution problems in this way may be termed the 'policy cycle'.

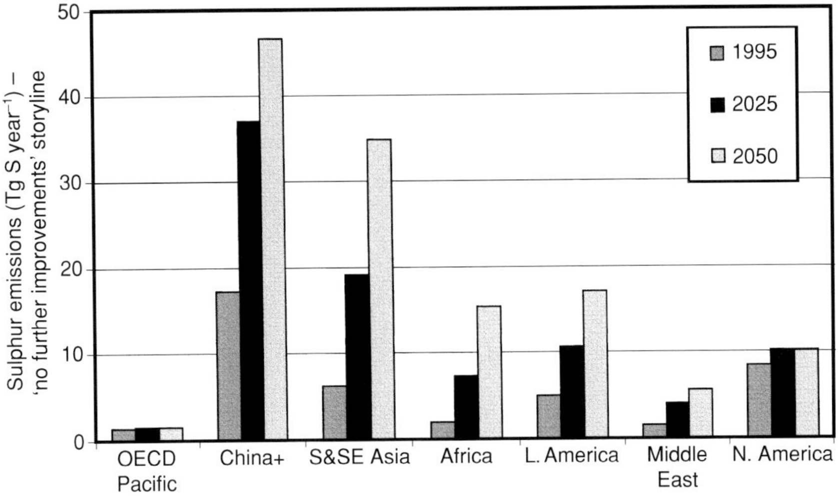

Fig. 12.1. Anthropogenic sulphur emissions for 1995 and those projected for 2025 and 2050 under the Global Scenario Group reference scenario (Raskin *et al.*, 1998) and assuming no improvements in sulphur emissions control beyond 1995 levels (i.e. the 'no further emission control' storyline). (Source: Kuylenstierna and Hicks, 1998.)

The policy cycle will not always take the same form but many aspects will be common to all regions attempting to solve air pollution problems.

In developing-country regions, policy initiatives at the regional scale have only started recently. Air pollution is emerging as a priority concern in many developing country regions, particularly in parts of Asia. Emissions of sulphur (Fig. 12.1), nitrogen and particulate matter are rising rapidly in many developing countries (Galloway, 1995). These emissions are leading to problems, particularly in urban areas and their fringes (e.g. Wahid *et al.*, 1995; Xu *et al.*, 1995). Problems are also being experienced at the regional scale. One graphic example of this is the haze over South-east Asia caused by forest fires in Indonesia (Chapters 1, 8 and 11) and another is the evidence for acidification in parts of China (Chapter 7). There is anxiety over the potential for transboundary pollution in parts of Asia and thus regional policy initiatives have begun. Projected emission increases in some areas suggest that the widespread damage experienced in Europe and North America could occur across widespread areas of Asia (Galloway, 1995). Countries developing their economies have an opportunity, in the planning phase, to avoid some of the problems associated with air pollution, as prevention is substantially cheaper than controlling emission sources once they have been built. Efficiency increases, leading to lower emissions through careful planning and investment, will also lead to greater economic efficiency within the countries.

This chapter will focus on the regions of East, South-east and South Asia, southern Africa and the MERCOSUR countries of Latin America and will explain the status and development of regional policy initiatives on air pollution. These regional/sub-regional activities are being developed through different intergovernmental organizations (e.g. Table 12.1 and Fig. 12.2) building, in some cases, on the experiences of intergovernmental cooperation in Europe. Facilitation of the policy process development has been achieved through a number of national and international programmes. One such programme is the Regional Air Pollution in Developing Countries (RAPIDC) Programme, which is managed by the Stockholm Environment Institute (SEI) and funded by the Swedish International Development Cooperation Agency (Sida). Activities in this Programme are used to illustrate the status of policy development in some regions.

2 Status of regional policy initiatives

Asia-Pacific sub-regions are in varying stages of the policy cycle. The sub-regional divisions in Asia used here are: East Asia, South-east Asia and South Asia.

2.1 East Asia

In the East Asian sub-region the Japanese Environmental Agency advocated an acidic deposition monitoring network in order to curb and prevent acid rain throughout the sub-region (EA, 1997). Four meetings have been held since 1993 to initiate the establishment of an acidic deposition monitoring network. Subsequently, the first intergovernmental meeting, held in March 1998, reviewed the draft documents for the design of the Acid Deposition Monitoring Network in East Asia (EANET) and adopted the implementation of preparatory-phase activities. Monitoring work, according to guidelines produced by EANET, has begun as part of these activities. The countries that agreed to participate are China, Indonesia, Japan, Republic of Korea, Malaysia, Mongolia, Philippines, Russian Federation, Thailand and Vietnam. The network is to be officially established in 2000, and will take into account the results of the preparatory-phase activities.

2.2 South-east Asia

In South-east Asia the issue of transboundary pollution was first highlighted in the 1990 Kuala Lumpur Accord on Environment and Development. The 1992 Singapore Summit identified transboundary pollution as among the major

Table 12.1. Examples of intergovernmental organizations involved in developing policy initiatives on regional air pollution in southern Africa, South America and South Asia.

Organization	Established	Members	Aim
ASEAN[a]	1967	Brunei Darussalam, Cambodia, Indonesia, Laos, Malaysia, Myanmar, Philippines, Singapore, Thailand, Vietnam	To encourage regional economic, social and political security, and cultural cooperation among the South-east Asian countries
MERCOSUR[b]	1991	Argentina, Brazil, Paraguay, Uruguay	To increase regional economic cooperation
SACEP[c]	1982	Afghanistan, Bangladesh, Bhutan, India, Iran, Maldives, Nepal, Pakistan, Sri Lanka	To promote and support the protection, management and enhancement of the environment, both natural and human, of the countries of South Asia, individually, collectively and cooperatively To make judicious use of the resources of the environment towards removal of poverty, reduction of socio-economic disparity, improve the quality of life and prosperity on a continuing basis To make the fullest use of the organizational arrangements and facilities for cooperation under SACEP
SADC[d]	1992	Angola, Botswana, Lesotho, Malawi, Mauritius, Mozambique, Namibia, South Africa, Swaziland, Tanzania, Zambia, Zimbabwe	To promote regional economic development and integration

[a]ASEAN, Association of South-East Asian Nations.
[b]MERCOSUR, Mercado Comun del Sur (Common Market of the South).
[c]SACEP, South Asia Cooperative Environment Programme.
[d]SADC, Southern African Development Community.

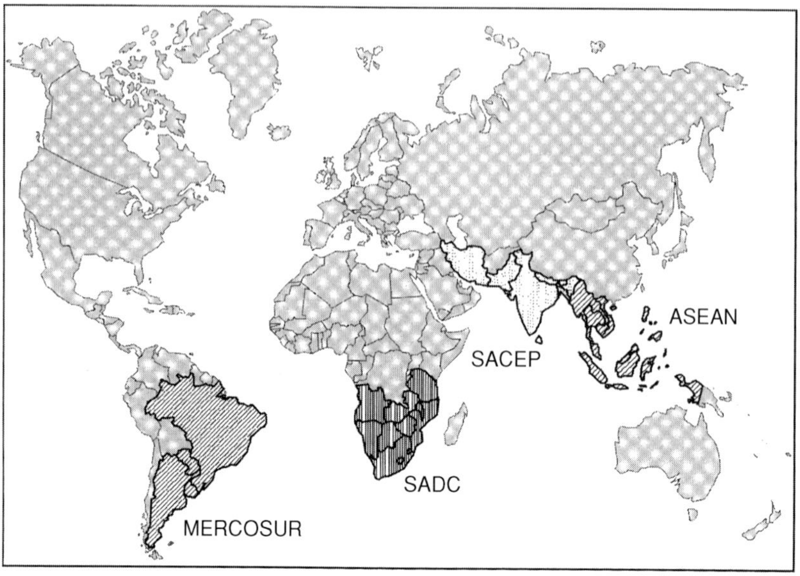

Fig. 12.2. Geographical location of intergovernmental organizations involved in developing policy initiatives on regional air pollution in southern Africa (SADC, Southern African Development Community), South America (MERCOSUR, Common Market of the South; Mercado Comun del Sur) and South Asia (ASEAN, Association of South-East Asian Nations; SACEP, South Asia Cooperative Environment Programme).

environmental concerns of the Association of South-East Asian Nations (ASEAN). The issue was addressed in the 1992 Singapore Resolution on Environment, and in the 1994 Bandar Seri Begawan Resolution on Environment and Development. An ASEAN meeting on the management of transboundary pollution was held in Kuala Lumpur in June 1995 and the ASEAN Cooperation Plan on Transboundary Pollution (ASEAN, 1995) was adopted. The ASEAN Cooperative Plan on Transboundary Pollution consists of three programme areas: transboundary atmospheric pollution; transboundary movement of hazardous waste; and transboundary shipborne pollution.

The programme area on transboundary atmospheric pollution has been set up: (i) to assess the origin and cause, nature and extent of local and regional haze incidents; (ii) to prevent and control the sources of haze at both national and regional levels by applying environmentally sound technologies and by strengthening both national and regional capabilities in the assessment, mitigation and management of haze; and (iii) to develop and implement national and regional emergency response plans.

The Haze Technical Task Force was established in 1995 to operationalize and implement the measures included in the ASEAN cooperation plan on transboundary pollution. However, an absence of specific operational plans rendered the cooperation plan ineffective. Following the 1997 fire-and-haze event, a Regional Haze Action Plan was formulated by the Haze Technical Task Force and was endorsed by the ASEAN Ministerial Meeting in 1997. The Regional Haze Action Plan has an operational focus, intending to identify specific actions to be taken at regional, sub-regional and national levels to prevent human-induced haze events.

2.3 South Asia

Of the Asian sub-regions, South Asia (including Bangladesh, India, Pakistan, Nepal, Iran, Bhutan, Maldives and Sri Lanka) is at the earliest stage of the policy cycle. In 1997 the United Nations Environment Programme/Environmental Assessment Programme for Asia and the Pacific (UNEP/EAP–AP) and SEI signed an agreement to create an Air Pollution Information Network and to promote regional cooperation in the region as part of the RAPIDC Programme funded by Sida. The approach employed by the RAPIDC Programme to develop regional cooperation in the South Asia region is illustrated in Fig. 12.3.

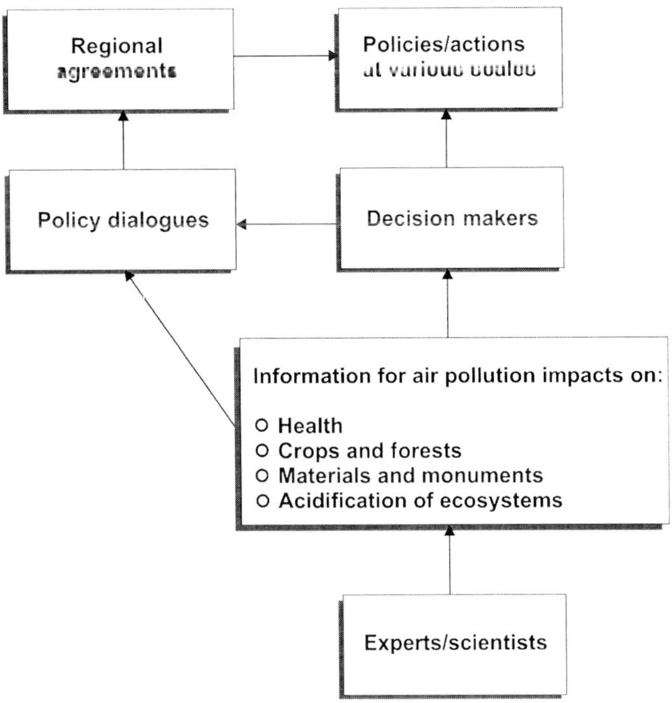

Fig. 12.3. Flowchart showing RAPIDC's approach to developing regional cooperation on air pollution control in the South Asia Region.

The first action of the initiative was to establish a network of experts and policy makers at national, sub-regional and regional levels (Fig. 12.4). The establishment of the network was based on the existing institutional base and networks of the UNEP/EAP–AP with the Ministry of Environment departments being the policy-level focal points at the national level. In addition to the ministries, national experts were also included as focal points responsible for carrying out scientific research at the national level. The intergovernmental organization, the South Asia Cooperative Environment Programme (SACEP) (see Table 12.1 and Fig. 12.2), which potentially would play a major role in formulation and implementation of sub-regional and national action plans, also formed an essential part of the network. In South Asia, SACEP has been set up purely for environmental reasons and has proved to be an important forum for the air pollution issue in the region. Recognizing the need for collaboration between the network and other major players on the issue, the Asian Development Bank (ADB) and ASEAN were also included in the network. Network establishment is a continuous process and additional institutions and experts, both national and international, are added regularly.

In order to 'fuel' regional cooperation and generally raise regional awareness of air pollution and its effects, SEI and UNEP/EAP–AP accessed all relevant and current information on the issue and produced a background document

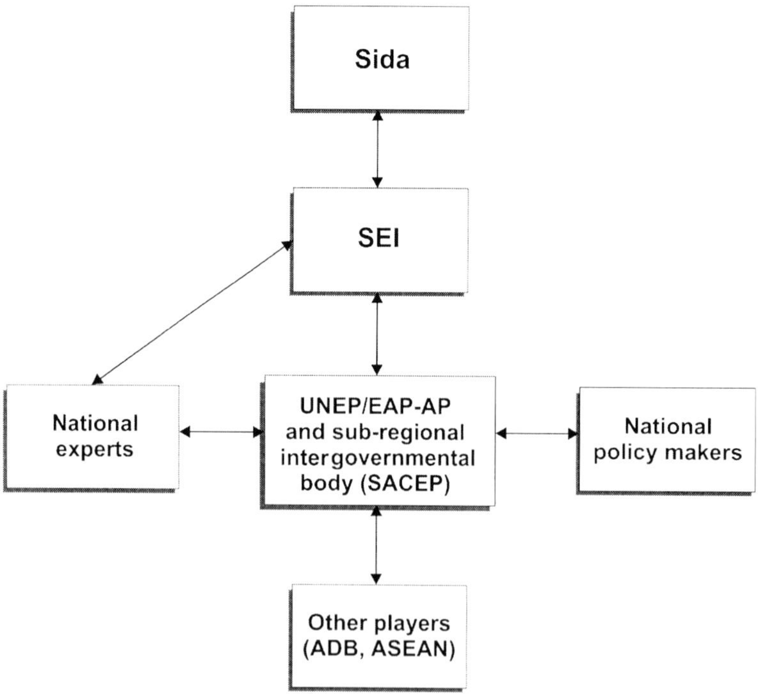

Fig. 12.4. Organization of the RAPIDC South Asia regional cooperation network.

(Kuylenstierna and Hicks, 1998). This was achieved by utilizing a team of national and international experts/scientists to collate the most up-to-date information on air pollution issues in the region. The background document included chapters on the regional perspective, the European experience, details of existing and potential impacts in the region, projections of the emission and deposition of pollution in the future (given current development patterns) and options for emission prevention and control. Information on the potential effects of air pollution was derived predominantly from research funded by the RAPIDC Programme. This research included activities such as the global assessment of ecosystem sensitivity to acidic deposition mapping (Cinderby *et al.*, 1998) and atmospheric pollution modelling (e.g. Robertson *et al.*, 1999) and monitoring (e.g. Granat *et al.*, 1996). An important aspect, which has been developed as part of RAPIDC, is the assessment of risks of air pollution impacts in the future, using Global Scenario Group projections (Raskin *et al.*, 1998).

Once the network was in place and the necessary information and documents were available, a round-table policy dialogue regarding the rapidly increasing problem of regional air pollution, with a focus on South Asia, was held in March 1998. UNEP/EAP–AP, based at the Asian Institute of Technology (AIT), Bangkok, Thailand, hosted the dialogue. The meeting was attended by a distinguished group of senior-level environmental ministry officials from South Asian countries, analysts and people with an influence on policy, and representatives of key environmental organizations in the region. The meeting agreed on the need for regional cooperation and, noting the experience of Europe, prepared a draft declaration. The declaration was approved in principle and was then submitted to the Governing Council of SACEP, a high-level body of environmental ministries of the regional countries, for approval.

The Malé Declaration

The seventh meeting of the Governing Council of SACEP was held in April 1998 in Malé, the Republic of Maldives, and adopted the declaration, naming it the 'Malé Declaration on Control and Prevention of Air Pollution and its likely Transboundary Effects for South Asia' (see Appendix 12.A). The Malé Declaration is an important step towards promoting intergovernmental cooperation for assembling information and expert knowledge related to regional air pollution.

The Malé Declaration stated the need for countries to carry forward, or initiate, studies and programmes on air pollution in each country of South Asia. The first stage in this process is to document current knowledge and information/institutional capacity in each nation relevant to air pollution issues. To this end it was agreed that baseline studies would be developed. Gaps in the current status of knowledge and capacity would become apparent and national action plans to fill these gaps could then be implemented, creating a solid scientific basis for the policy process. Implementation of the action plan will put in place expertise, equipment and information for quantitative monitoring, analysis and policy recommendations for eventual prevention of air pollution (Fig. 12.5).

A Memorandum of Understanding (MoU) on working arrangements between the SEI and the UNEP/EAP–AP in the implementation of the Regional Air Pollution Programme 'Malé Declaration on Control and Prevention of Air Pollution and its likely Transboundary Effects for South Asia' was signed in May 1998. National Implementing Agencies (NIAs) were nominated by the Ministry of Environment of each country to carry out the national baseline studies and action plan. An inception workshop took place in February 1999 and implementation plans and guidelines produced were agreed at the meeting. These formed the templates for ongoing activities being carried out by the NIAs in each South Asian country to develop the baseline studies and action plans.

Following the Malé Declaration, a meeting was held in Hyderabad, India (August 1999), on air pollution and health issues in rapidly developing

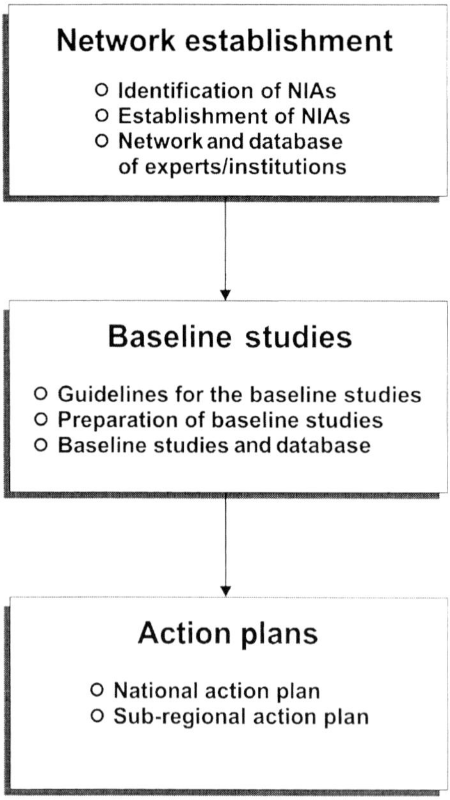

Fig. 12.5. Flowchart showing the scheme for the implementation of the Malé Declaration by the National Implementing Agencies (NIAs).

countries (McGranahan and Murray, 1999). The meeting had a South Asian focus and was attended by health experts, policy makers, NIAs and representatives of international and intergovernmental organizations. This type of science/policy-maker dialogue, related to a specific issue, will support the development of the Malé Declaration process.

2.4 Latin America

Air pollution problems in Latin America are not as widespread as in Asia and, as a consequence, are not as high on political agendas, except in the 'megacities'. Some of these cities are amongst the most polluted in the world. In urban environments air pollution has been considered a serious problem for many years (Rodhe *et al.*, 1988) and emissions of sulphur dioxide, nitrogen oxides and ammonia and concentrations of tropospheric ozone have been increasing over the last few decades (Galloway, 1989; Sanhueza, 1997). Only when large point sources of pollution, such as power stations, are situated near national boundaries does the pollution result in a transboundary problem. However, as emissions increase, the problem will become increasingly regional.

The air pollution issue has been addressed at the local scale in cities, such as Mexico City, where certain air pollution control measures have been implemented imposing restricted use of motor vehicles, temporary closure of industries and policies to move industries away from cities (Chapter 3). Until recently, regional initiatives in South America have been limited to activities associated with the climate change debate and the Kyoto Protocol, but there are now activities occurring with regard to regional air pollution.

In view of the projected increases in pollutant emissions for the next century, and the possibility of a more regional problem, a second RAPIDC regional policy dialogue was held in Buenos Aires, Argentina, in June 1998. The dialogue successfully initiated regional discussions on air pollution in the MERCOSUR (Common Market of the South) countries (see Table 12.1 and Fig. 12.2) of South America. The main outcome was the Cañuelas Declaration (see Appendix 12.B). The meeting was organized for country members of the MERCOSUR Free Trade Agreement by Fundación Futuro Latinoamericano (FFLA) in Ecuador, the International Institute of Environmental Development for Latin America (IIED-LA) in Argentina, and by SEI. The main aim of the policy dialogue in Buenos Aires was to analyse and discuss among different stakeholders the importance of air pollution within the MERCOSUR countries, and to seek appropriate ways to orientate and influence public policies to tackle the problem. Policy makers and leaders of different stakeholder communities from the member countries of the MERCOSUR Free Trade Agreement (Argentina, Brazil, Uruguay and Paraguay) participated during the 2-day meeting. In particular, there was representation from industrial unions and associations, non-governmental organizations (NGOs), academia, government and the media.

Although air pollution and climate change issues are strongly related to technology, a special focus was directed at the analysis of possible political responses and the ways to support the decision-making process. From such a perspective, the delegates concentrated on formulating proposals to be presented to high-level policy fora, regional country governments and MERCOSUR treaty organizations. The Cañuelas group (signatories to the Cañuelas Declaration) partners view the process of identifying air pollution sources and their relationship with environmental problems within the region as complementary to studies associated with COP5 (5th Conference of Parties to the Framework Convention on Climate Change (FCCC) for the Kyoto Protocol) and the climate change debate.

In Latin America the focus on MERCOSUR is useful as it provides an increasingly strong intergovernmental organization in Latin America. Air pollution is not seen as a transboundary problem, but more as a shared problem in the region. The focus then has been more on trade issues and on the harmonization of legal frameworks to ensure that there is a level playing field for industry with respect to environmental regulations. This also forms the focus of the follow-up to the Cañuelas Declaration. A document, agreed by the Cañuelas Group, has been produced concerning the need for harmonization of legal frameworks for the prevention of air pollution in MERCUSOR countries. This document was instigated to coincide with the development of an Environmental Protocol for the MERCOSUR Free Trade Agreement, which is to include the harmonization of policies and mechanisms as a first step. The document will be presented to the Environment Protocol Ratification Committee, with the objective of influencing the Environmental Protocol discussion at the next MERCOSUR high-level policy forum. IIED-LA is coordinating this activity with the National Secretariat for Natural Resources and Sustainable Development (Argentina) and the Cañuelas Group.

2.5 Southern Africa

Southern Africa is in a process of land use transformation similar to that which took place in Europe and North America earlier this century. Industrial emissions of air pollutants in southern Africa only started in the mid-20th century and impacts are not yet as serious as those that have occurred in the northern hemisphere. The past few decades have seen the region transform itself from a rural society to a complex one that has made great strides in industrialization, urbanization and economic development. Industries such as agriculture, mining, forestry and manufacturing, together with a growing population, have combined to bring about some environmental problems that were virtually non-existent in the southern African region at the turn of the 20th century.

While many of the Southern African Development Community (SADC) countries acknowledge that air pollution is a problem, only South Africa and Zambia are seen as the major contributors to industrial air pollution. Practices such as domestic fuel burning and grassland burning for grazing are widespread in the region, but the impacts on human and ecosystem health are not apparent. In terms of atmospheric pollution, the main issues in policy initiatives are on ozone-depleting substances and greenhouse gases (driven by the global process). Virtually all SADC member states have policies on air pollution control in one form or another. These have been covered in either National Conservation Strategies (NCS), National Environmental Action Plans (NEAPs) or a new generation of environmental laws. While the necessary laws and policies may be in place, their implementation is lacking, with the result that there is little or no regulation of air pollution sources.

The majority of SADC countries have ratified both the 1987 Montreal Protocol on Substances that Deplete the Ozone Layer and the 1992 Framework Convention on Climate Change. While this shows the commitment by SADC countries to global issues, the immediate problems of air pollution do not seem to have had similar attention. Action plans exist for the phase-out of ozone-depleting substances and reduction of greenhouse gases but no such plans exist for air pollution in general.

Regional initiatives do exist such as the SADC Treaty on 'Policy and Strategy on Environment and Sustainable Development', which acknowledges transboundary air pollution as a major policy issue. However, little seems to have been done in terms of policy initiatives. Indications are that air pollution has not received sufficient attention, either at national or regional levels, at least when compared with other environmental initiatives.

Under the auspices of the Southern African Development Community Environment and Land Management Sector (SADC-ELMS) (see Table 12.1 and Fig. 12.2), a third RAPIDC regional policy dialogue meeting was convened on 29–30 September 1998 in Harare, Zimbabwe, on 'Regional Air Pollution Issues in Southern Africa'. Participants at the workshop were policy makers drawn from various national ministries such as health, environment, natural resources, mines and energy. In addition, representatives from industrial, scientific and research communities attended. The focus of the meeting was to initiate a policy dialogue on the problem of regional and transboundary air pollution in Africa. The meeting was opened by the Hon. S.K. Moyo, Minister of Mines, Environment and Tourism of Zimbabwe. The output from the meeting was 'The Harare Resolution on the Prevention and Control of Regional Air Pollution in Southern Africa and its Likely Transboundary Effects' (see Appendix 12.C). The industries present at the meeting were supportive of the process – for example, the power sector (ESKOM) and the mining sector (from Zambia).

As was the case in South Asia, the RAPIDC Programme aided regional cooperation in southern Africa by establishing a regional network, which became known as the Air Pollution Information Network for Africa (APINA). APINA is currently in the process of being officially recognized by SADC as a group that will be able to help to develop protocols relating to air pollution in the SADC region. In 2000, SADC-ELMS is due to develop an Environmental Charter which, when signed by the environment ministers of each country, will allow SADC-ELMS to develop protocols on different environmental issues. Apart from being well placed to aid protocol development with SADC-ELMS, APINA is gaining recognition in southern Africa as a link between science and policy making. APINA would have a role to play in understanding the policy implications of the findings of research programmes under way in the region. An example of such a programme is SAFARI 2000 – a southern African regional science initiative which involves numerous international collaborators.

3 Summary and conclusions

Air pollution is an issue traditionally associated with the industrialized countries of Europe and North America and not so readily with developing countries. However, such has been the pace of industrialization in some of these countries, coupled with a lack of access to clean and modern technology due to financial or other reasons, that the air pollution issue is gaining in importance in many regions. In some countries, air pollution has been tackled as a local issue, but in this chapter activities have been reviewed at a regional scale for two reasons: firstly, by necessity transboundary issues occur at a regional scale and, secondly, dealing with air pollution and other environmental issues at this scale has been very successful in Europe. In Europe, regional activities in preparation for the negotiations have been of great value as they provide a forum for international exchange of relevant information among experts who often advise national governments. In addition, activities underpinning negotiation may directly or indirectly impact significantly on national decision making when it comes to taking measures for reducing or preventing emissions.

The initiation of international cooperation in developing country regions can be triggered by dramatic impacts at regional level (such as the Indonesian haze episode), or by increasing awareness in the policy-making community of an escalating problem. Increased contact between international and regional expert knowledge with the policy community is considered to be one of the crucial ingredients for developing the motivation to address regional air pollution problems.

In this chapter the emphasis has been on the development of agreements between national governments and their agencies, but there are other important stakeholders in the process leading to prevention and control of air pollution that are not explicitly mentioned. These include different sectors of the public, industry and NGOs. International agreements can provide a useful framework within which these and other stakeholders can participate in air pollution management.

The status of the policy cycle differs from region to region. In regions where air pollution emissions are already high the degree of international cooperation is greater. In some regions the problems are clearly transboundary in nature. The haze episode in ASEAN countries was the most obvious example of this. The political implications have already resulted in a haze action plan, with regular meetings at both expert and ministerial level. In East Asia there is increasing concern about the potential for transboundary transport that has resulted in cooperative monitoring of air pollution (EANET). In South Asia there is little evidence of the air pollution problem being transboundary in nature, although the information needed to demonstrate the magnitude of the current problem is not available. There is, however, concern that this is an important issue that has the potential to increase rapidly in the region. Therefore, there is interest and agreement within the region concerning the need to collaborate on issues related to air pollution. Countries of South Asia have agreed to carry out studies as part of the implementation of the Malé Declaration to determine the current and potential future status of air pollution. The studies lay the foundation for collaboration and sharing of experience, knowledge and technology that exist in the region which, in turn, will enable the creation of harmonized and coordinated activities related to, for example, monitoring, impacts, emission inventories, standards or legislation.

In many developing country regions, urban air pollution or pollution around heavy industrial concentrations is the main concern. In the MERCOSUR countries there is strong support for dealing with the issue of air pollution in a regional manner to ensure that legal frameworks are harmonized within this free-trade region, and also to develop information concerning this shared problem. In southern Africa, the anticipated emission increases are of considerable concern but there is the will to learn from regions where pollution damage has already occurred. There is also a general wish to prevent air pollution damage in the region, particularly as the countries of southern Africa have signed global treaties concerning climate change and ozone depletion and it is considered timely to deal with regional air pollution issues as either a shared or transboundary problem. In Latin America and Africa the urgency is not as great as in Asia, where air pollution forms a large and growing problem.

Europe has been damaged by a century or more of heavy pollution. Impacts have been caused to urban environments, crops and natural ecosystems and now agreements are attempting to limit emissions and reduce the damage caused. Developing countries, whilst building their industrial base and infrastructure, have an opportunity to learn from this experience and 'leap-frog' to a situation where well-being of their populations is increased whilst avoiding large emission increases and the damage associated with these emissions. Expertise developed in Europe can be used to facilitate this desired situation. Governments and the donor community have an important role in facilitating the development and transfer of expert knowledge and also the experience of international cooperation in order to manage the problems and levels of air pollution.

Acknowledgements

Grateful thanks to Barnabas Chipindu (University of Zimbabwe), Laurent Mermet (ENGREF, France), John Innes and Isobel Devane for useful comments on the manuscript and to Erik Willis and Harry Vallack for help with graphics.

References

ASEAN (1995) *ASEAN Co-operative Plan on Transboundary Pollution.* ASEAN Secretariat, Jakarta.

Battarbee, R.W. (1990) The causes of lake acidification, with special reference to the role of acid deposition. *Philosophical Transactions of the Royal Society of London, Series B* 327, 339–347.

Bull, K. and Fenech, G. (1999) International activities to reduce pollution impacts at the regional scale. In: Innes, J.L. and Oleksyn, J. (eds) *Forest Dynamics in Heavily Polluted Regions.* CAB International, Wallingford, UK, and IUFRO, Vienna, pp. 219–230.

Cinderby, S., Cambridge, H.M., Herrera, R., Hicks, W.K., Kuylenstierna, J.C.I., Murray, F. and Olbrich, K.A. (1998) *Global Assessment of Ecosystem Sensitivity to Acidic Deposition – Version 2, April 1998.* Stockholm Environment Institute, York, UK.

EA (1997) *Acid Deposition Monitoring Network in East Asia – Achievements of the Expert Meetings.* Environmental Agency of Japan, Tokyo.

Elder, F.C. and Brydges, T. (1983) *Effects of Acid Precipitation on Aquatic Regimes of Canada.* The Canadian contribution to the Committee on the Challenges of Modern Society. NATO Panel 3 Report. NATO, Brussels.

Galloway, J.N. (1989) Atmospheric acidification: projections for the future. *Ambio* 18, 161–166.

Galloway, J.N. (1995) Acid deposition: perspectives in time and space. *Water, Air, and Soil Pollution* 85, 15–24.

Granat, L., Suksomsankh, K., Simachaya, S., Tabucanon, M. and Rodhe, H. (1996) Regional background acidity and chemical composition of precipitation in Thailand. *Atmospheric Environment* 30, 1589–1596.

Kuylenstierna, J. and Hicks, W.K. (eds) (1998) *Regional Air Pollution in Developing Countries, Background Document for Policy Dialogue, Bangkok, March 1998.* Stockholm Environment Institute, York, UK.

McGranahan, G. and Murray, F. (1999) *Health and Air Pollution in Rapidly Developing Countries.* Stockholm Environment Institute, York, UK.

Raskin, P., Gallopin, G., Gutman, P., Hammond, A. and Swart, R. (1998) *Bending the Curve Toward Global Sustainability. A Report of the Global Scenario Group.* PoleStar Series Report No. 8. Stockholm Environment Institute, Boston, Massachusetts.

Robertson, L., Langner, J. and Engardt, M. (1999) An Eulerian limited-area atmospheric transport model. *Journal of Applied Meteorology* 38, 190–210.

Rodhe, H., Cowling, E., Galbally, I., Galloway, J. and Herrera, R. (1988) Acidification and regional air pollution in the tropics. In: Rodhe, H. and Herrera, R. (eds) *Acidification in Tropical Countries.* SCOPE 36. John Wiley & Sons, Chichester, UK, pp. 3–39.

Sanhueza, E. (1997) Deposition in South and Central America. In: Whelpdale, D.M. and Kaiser, M.S. (eds) *Global Acid Deposition Assessment.* Global Atmospheric Watch Publication No. 106. World Meteorological Office, Geneva, pp. 135–144.

UN-ECE (1996) *1979 Convention on Long-range Transboundary Air Pollution and its Protocols.* ECE/EB.AIR/50. United Nations Economic Commission for Europe, Geneva, Switzerland.

Wahid, A., Maggs, R., Shamsi, S.R.A., Bell, J.N.B. and Ashmore, M.R. (1995) Effects of air pollution on rice yield in the Pakistan Punjab. *Environmental Pollution* 90, 323–329.

Xu, X., Dockery, D.W., Christiani, D.C., Li, B. and Huang, H. (1995) Association of air pollution with hospital outpatient visits in Beijing. *Archives of Environmental Health* 50, 214–220.

Appendix 12.A: Malé Declaration on Control and Prevention of Air Pollution and its Likely Transboundary Effects for South Asia

Recognizing the potential for increase in air pollution and consequential phenomena due to concentrations of pollutant gases, acid rain or acid deposition as well as the impacts on the health of humans and other living organisms in all our countries due to man made and natural causes;

recognizing the potential for increase in transboundary air pollution as a corollary of air pollution in each country;

realizing that the potential for air pollution increase and its transboundary effects will accumulate in the absence of national measures to abate and prevent such potential;

reiterating in this context Principle 21 of the UN Declaration on the Human Environment in 1972 which stated that: 'states have, in accordance with the charter of the United Nations and the principle of international laws, the sovereign right to exploit their own resources pursuant to their own environmental policies, and the responsibility to ensure that activities within their jurisdiction or control do not cause damage to the environment of other states or of areas beyond the limits of national jurisdiction';

keeping in mind the need for constant study and monitoring of the trends in air pollution with a view to understanding the extent of our potential for damage to the environment and health in the member countries and taking consequential measures to strengthen and build capacity for such activities;

stressing the need for development and economic growth that will help build up the quality of life and incomes of all the people of all the region, in particular the poorer sections of the population, having due regard to the need to have a clean and healthy environment;

emphasizing that air pollution issues have to be analysed and managed in the wider framework of human and sustainable development within each country and within the region; and

drawing from the experience of cooperation in the region in matters such as cultural exchange and also from the experience in other regions such as Europe and sub-regions of Asia, for example, ASEAN and East Asia.

We declare that countries of this region will initiate and/or carry forward programmes in each country to:

1. Assess and analyse the origin and causes, nature, extent and effects of local and regional air pollution, using the in-house capability in identified institutions, universities, colleges, etc., building up or enhancing capacities in them where required;

2. Develop and/or adopt strategies to prevent and minimize air pollution;

3. Work in cooperation with each other to set up monitoring arrangements beginning with the study of sulphur and nitrogen and volatile organic compounds emissions, concentrations and deposition;

4. Cooperate in building up standardized methodologies to monitor phenomena such as acid depositions and analyse their impacts without prejudice to the national activities in such fields;

5. Take up the aforesaid programmes and training programmes which involves the transfer of financial resources and technology and work towards securing incremental assistance from bilateral and multilateral sources;

6. Encourage economic analysis that will help to arrive at optimal results;

7. Engage other key stakeholders, for example industry, academic institutions, NGOs, communities and media, etc., in the effort and activities.

We also declare that we shall constantly endeavour to improve national reporting systems and strengthen scientific and academic effort in the understanding and tackling of air pollution issues.

We further declare that we shall continue this process in stages with mutual consultation to draw up and implement national and regional action plans and protocols based on a fuller understanding of transboundary air pollution issues.

We declare that in pursuit of the above, we shall evolve, as appropriate, institutional structures at the national level, including networking, both for the purposes of policy and the technical requirements, and we shall use the good offices of regional, international bilateral and multilateral agencies in this, as appropriate.

Malé, the Maldives
April 1998

Appendix 12.B: The Cañuelas Declaration on the Control and Prevention of Atmospheric Pollution (in MERCOSUR countries)

a. Declaration of Principles:

1. The group that met at Cañuelas recognizes that atmospheric pollution is a real and important problem in the MERCOSUR countries, directly affecting the population and the environment.
2. The existence of MERCOSUR opens up new opportunities for the continuation of work on this issue, within a broad and regional framework. The existence of Working Sub-group No. 6 within MERCOSUR, and of the proposal for an Environment Protocol approved by the sub-group, and the consultative economic and social forum, form the institutional bases for the generation of new proposals.
3. In the region efforts to control and prevent the worsening of atmospheric pollution are being carried out. Nevertheless, it is critical to strengthen, broaden and deepen the actions.
4. The group is aware that the responses have to come both from individual commitments of all members of society as well as from institutions.
5. The existence of MERCOSUR facilitates cooperation, information and experience exchange, and exchange of initiatives concerning issues of atmospheric pollution. Given that, it is important to take into account the following:

- that the established framework should stimulate the citizens to contribute to the actions of prevention;
- that the actions of prevention, mitigation or restoration should be carried out based on studies that justify such actions;
- that the solutions should correspond to the knowledge of local and regional conditions;
- that alternative scenarios should be developed to identify trends and options to guide development policies consistent with sustainability.

6. The energy supply, transport, industry, residential and agriculture are priority sectors where concrete, viable and fundable efforts should be concentrated to solve the problems of air pollution.
7. The group believes that it is important to promote and increase the discussion of these issues in their own constituencies in order to increase awareness, promote solutions and actively participate in the proposed actions.

b. Action Plan:

The commitments and responsibilities should be shared among the different sectors. Those present at the meeting have promised the following:

1. *Environmental studies*: activities will be promoted such that governments, with the support of the academic sector, NGOs and other sectors that could contribute to this effort, will promote the development of emission inventories for atmospheric pollution and their relation to environmental problems in the region. SEI will contribute by strengthening the Latin American Network regarding this subject as well as through the development and transfer of risk analysis methodologies.

The industrial sector, through cooperation amongst industrial federations in the four MERCOSUR countries will contribute to the implementation of emission inventories; a process that can begin immediately.

2. *Dissemination*: the media should be involved. The creation of a network of environmental journalists in the MERCOSUR region will be promoted in order to disseminate issues concerning atmospheric pollution issues, based on the existing network in Brazil.

Industrial associations are willing to collaborate in this initiative. They will increase their efforts, by broadening their database to include subjects related to atmospheric pollution together, with the media network.

The labour unions will participate in the dissemination of information to their affiliates.

3. *Technology*: chambers of industry will establish a network of centres to develop, transfer and disseminate clean and efficient technologies. Academia will be integrated into this network.

4. *Harmonization of legal frameworks*: governments, in collaboration with public ministries and NGOs, will promote awareness of the need to develop harmonized legal frameworks for issues relating to environmental pollution, among the relevant authorities.

IIED-LA will prepare a draft proposal for approval by the group which will subsequently be put forward for consideration of the Ministers of the Environment of the four MERCOSUR countries.

NGOs will participate in this process as facilitators of the dialogue and as sources of technical information.

In 1998 the MERCOSUR Governments are anticipated to ratify the environmental protocol which should include the harmonization of legal principles.

5. *Institutional development*: governments, workers' unions and NGOs at the Cañuelas meeting express their interests in supporting the process of institutional strengthening on the subject of atmospheric pollution. The NGOs propose the implementation of rapid assessment of needs for capacity building in different institutions.

Cañuelas, Province of Buenos Aires, Argentina
12 June 1998

Appendix 12.C: The Harare Resolution on the Prevention and Control of Regional Air Pollution in Southern Africa and its Likely Transboundary Effects

Participants of the Harare policy dialogue:

Acknowledging concern in the region that air pollution and its potential increase in Southern Africa is an issue, which demands more attention,

Reiterating the provisions of Agenda 21, Principle 2, which states that: 'States have, in accordance with the Charter of the United Nations and the principles of international law, the sovereign right to exploit their resources pursuant to their own environmental and developmental policies, and their responsibility to ensure that activities within their jurisdiction or control do not cause damage to the environment or other states or of areas beyond the limits of national jurisdiction',

Noting the existence of the Southern African Development Community (SADC) Policy and Strategy for Environment and Sustainable Development, which highlights the problem of transboundary air pollution,

Recalling the opportunity afforded by existing regional cooperation through the SADC Environment and Land Management Sector (SADC-ELMS) and its mandate to coordinate regional initiatives on environment and land management,

Recognizing the following impacts on:

- human health in southern Africa with associated economic consequences, particularly to the less privileged in society and to government health services;
- managed ecosystems on which people in southern Africa depend, involving agriculture, forestry, fisheries and wildlife utilization;
- cultural heritage, materials, buildings and infrastructure through corrosion, which may cause large economic losses due to increased maintenance costs and loss of value;
- climate change and biodiversity in southern Africa, including important flora and fauna of the region;
- forests, woodlands and savannah which may lead to localized vegetation loss, land degradation and even desertification,

Recognizing also the following constraints:

- that there is limited capacity in terms of funding, equipment and human resources to measure, assess, control and mitigate air pollution;
- that information on air pollution and its impacts is inadequate, often inaccessible and diffuse;

- that, as a result of limited public awareness, there is currently inadequate concerted pressure, action and participation to influence policy and practice on air pollution;
- that policy and legislation on air pollution, both at national and regional level, are not harmonized, and that enforcement of existing legislation is limited;
- that the potential for conflict exists between economic development and environmental protection both at national and regional levels,

Recognizing further the following opportunities:

- the intent of countries in the region to develop sustainably increased public, media and business awareness of air pollution and climate change;
- early awareness of these issues enables countries to consider development options and to acquire the necessary information on which to make informed decisions about investment in clean development technologies;
- current cooperation within SADC countries provides an opportunity for harmonizing legislation and other policy instruments on air pollution within the region,

Recalling the international agreements that have successfully addressed air pollution in industrialized regions, such as the United Nations Economic Commission for Europe (UN-ECE) Convention on Long-range Transboundary Air Pollution (LRTAP), which might provide useful guidance for developing appropriate legislation for Southern Africa,

Noting examples of existing legislation in southern Africa that have resulted in some improvement in air quality, there are opportunities for wider rationalization and effective application of such legislation,

Conscious of the existence, internationally, of a range of cleaner production technologies and efficient end-use technologies that reduce emissions,

Being aware of the availability of international financial mechanisms to support the transfer and sharing of technologies including those promoting cleaner production,

Recognizing that current initiatives within the southern African power pool provide opportunities for reductions in regional air pollution,

Recognizing also the availability of regional scientific and technical expertise within universities, research institutions, industry and government, and the considerable opportunity for further collaboration among themselves,

Recognizing further the importance of achieving the following goal: 'to reduce the incidence and the impact of air pollution in Southern Africa without compromising regional economic development',

We hereby *resolve* to request the SADC Council of Ministers, through SADC-ELMS, to develop a Protocol on Regional Air Quality and Atmospheric Emissions, taking into account the following issues:

- The need for harmonized and strengthened legislation.
- The importance of creating appropriate incentive structures.
- The advantages for developing strategies for increasing awareness and education.
- The benefits accruing from improved information availability and accessibility to information that can promote understanding.
- The importance of encouraging the use and development of improved technologies.
- The need for further cooperation to enhance regional capacity to assess and analyse the origin and causes, nature, extent and effects of local and regional air pollution, using the in-house expertise in identified institutions, universities, colleges, etc.
- The need to identify appropriate financial resources required to carry out the programmes, strategies and projects.
- The importance of developing, through SADC ELMS, a structure with required linkages in Southern Africa, to carry out coordinated programmes in building up and applying standardized methodologies to monitor emissions, concentrations, depositions and impacts.
- The importance of engaging stakeholders (industry, academic institutions, NGOs, communities, media, etc.) in these efforts and activities.

We further *resolve* that we shall continue to assist the policy process through mutual consultation in efforts to implement national and regional action plans and protocols based on an improved understanding of transboundary air pollution issues.

Harare, Zimbabwe
September 1998

Conclusions 13

A.H. Haron[1] and J.L. Innes[2]

[1]Forestry Department Peninsular Malaysia, Kuala Lumpur, Malaysia; [2]Department of Forest Resources Management, University of British Columbia, Vancouver, Canada

1 Problem identification

The chapters in this book indicate that air pollution is a significant environmental problem for rapidly industrializing countries and regions. Some countries, particularly those in Asia, are relatively advanced in the extent to which they recognize the problem. A considerable amount of work has been done on pollution impacts in India, Korea and China, but relatively little information is available for other areas. In many countries, there is a lack of awareness or even a lack of interest in the problem (Instituto Internacional de Medio Ambiente y Desarrollo – America Latina, 1992). Environmental pollution is often seen simply as a necessary cost of development. The pollution is derived from a number of sources, but some of the greatest problems are associated with forest fires. This contrasts markedly with developed countries in the temperate zone, where large-scale forest fires are not a major issue. However, as income levels rise, industrial pollution is likely to increase.

The main problems (not in order of importance) that can be identified today are:

- suspended particulates and other pollutants in smoke from forest fires;
- suspended particulates and other pollutants in smoke from domestic biomass burning;
- sulphur dioxide from coal-burning power plants and coal fires;
- fluorides from fertilizer plants, brick works, aluminium smelters and other sources;

- suspended particulates from a range of industrial sources, including automobile emissions and mining;
- heavy metals, primarily from automobile emissions;
- ozone, with the precursors being derived from biomass burning, automobiles and industry; and
- pesticide and herbicide residues.

2 Progress to date

In Chapter 12, a number of initiatives are described that outline some of the regional agreements that have been reached on environmental pollution. These are extremely encouraging, and demonstrate that there is a willingness to address the problem of air pollution in most rapidly industrializing countries. However, such agreements are only a first step. They need to be converted into national and international legislation, and then the legislation has to be enforced. Much the same can be said of emerging cooperative environmental management plans such as the North-east Pacific Action Plan and the North-east Asian Environment Programme (Hayes and Zarsky, 1994). These need to consider air pollution within the broader range of the suite of environmental problems facing a region.

Our understanding of some of the transboundary issues is growing rapidly. Models now exist that enable, for example, critical loads to be mapped over very large areas (Hettelingh *et al.*, 1995; see also Chapter 6). In some cases, these models differ from those used in Europe, such as the Asian version of the Regional Air pollution Information and Simulation model (RAINS–Asia); in others, the lack of reliable information on vegetation patterns and air quality limits the effectiveness of the models.

3 Recommendations

Given the present economic and social limitations that many countries face, efforts need to concentrate first on the achievement of air quality standards that protect the health of the human population. This is not an issue of science. The knowledge exists about effects and the technology exists for pollution abatement. Rather, it is an issue of costs and policy priorities. Once human health standards have been achieved, secondary standards can be developed to protect the environment. The adoption of air quality standards in themselves represents a challenge, as they need to be met without compromising economic growth and without compromising more pressing issues such as the reduction of poverty.

A major problem in gathering the information for this volume was the availability of information. In many countries, no information is collected.

There are very few fully equipped and functional air quality monitoring stations in the tropics, and the numbers in other areas covered by this volume is limited. Where they exist, access to the information has been a problem. Information access is no longer a technical issue as a result of the increasing availability of Internet-based information. However, it is a political issue, and in some jurisdictions there is an unwillingness to make air quality information generally available.

A related issue is the enforcement of air quality legislation. In some jurisdictions, air quality standards exist in law but there are no mechanisms to enforce them. In some cases, economically important but polluting industries are permitted to function because of the financial benefits they bring (either to the administration or to the population in general). Various steps are needed to address this problem, including the eradication of corruption in government administrations, the creation of adequately funded environmental protection agencies and the enforcement of international standards for air quality. Such standards have already been set by agencies such as the World Health Organization; the problem lies in meeting them, and knowing that they have been met. Two approaches exist: a market-based policy relying on the optimal allocation of environmental resources; and a command-and-control approach relying on the trade-off between environmental quality and economic progress (James, 1994).

With the exception of the fire-generated regional hazes, all the pollutants that are currently problems in rapidly industrializing countries were also problems for the developed countries. As a result, they have been intensively researched. While their effects on all species may not be known, their effects on humans are generally well understood, and the impacts on vegetation can be judged from what is already known.

There is a major lack of understanding on the impacts of current levels of pollution on the vegetation in rapidly industrializing countries. This knowledge will increase slowly, but as long as human health is jeopardized, research into vegetation effects will take second place. Such research will be difficult, as it is hindered by the poor knowledge of ecosystem functioning in many of the countries covered by this report, particularly in tropical Africa. The International Union of Forest Research Organizations could play a significant role in developing research into the interactions between forests and air pollution in rapidly developing countries. In this respect, it is unfortunate that there is no research group specifically dealing with air pollution problems in rapidly industrializing countries.

One of the issues emerging from the various contributions to this book is the importance of interdisciplinary research. The problems facing the adoption of improved cooking stoves illustrates this well. Even when there is a technological solution, a lack of appropriate socio-economic research may hinder its adoption. In the case of cooking stoves, this has happened repeatedly, indicating the need to integrate social and economic research with research into impacts and new technology. It is an issue that appears to be particularly important in developing countries, where technological solutions are advocated by aid agencies without a full assessment of the factors that might influence the acceptance of the solution. Consequently, any final recommendation that we make must be for the development of solutions to environmental problems that integrate the environmental, economic and social needs of the people affected by the problem.

References

Hayes, P. and Zarsky, L. (1994) Environmental issues and regimes in North-east Asia. *International Environment Affairs* 6, 283–319.

Hettelingh, J.P., Sverdrup, H. and Zhao, D.W. (1995) Deriving critical loads for Asia. *Water, Air, and Soil Pollution* 85, 2565–2570.

James, V.U. (1994) The link between poverty and environmental degradation in sub-Saharan Africa. In: James, V.U. (ed.) *Environmental and Economic Dilemmas of Developing Countries: Africa in the Twenty-first Century*. Praeger, Westport, Connecticut, pp. 69–80.

Instituto Internacional de Medio Ambiente y Desarrollo – America Latina (1992) Sustainable development in Argentina. *Environment and Urbanization* 4, 37–52.

Appendix A: Conversion Factors

Conversion of pH values to microequivalents hydrogen ion per litre

$\mu eq\ H^+\ l^{-1} = antilog\ (6.0 - pH)$

pH	$\mu eq\ H^+\ l^{-1}$	pH	$\mu eq\ H^+\ l^{-1}$	pH	$\mu eq\ H^+\ l^{-1}$
3.0	1000	4.0	100	5.0	10
3.1	794	4.1	79	5.1	8
3.2	631	4.2	63	5.2	6
3.3	501	4.3	50	5.3	5
3.4	398	4.4	40	5.4	4
3.5	316	4.5	32	5.5	3
3.6	251	4.6	25	5.6	3
3.7	200	4.7	20	5.7	2
3.8	158	4.8	16	5.8	2
3.9	126	4.9	13	5.9	1

Units for concentration in precipitation

Ammonium: $1\ \mu mol = 1\ \mu eq\ l^{-1} = 0.014\ mg\ N\ l^{-1} = 0.018\ mg\ NH_4^+\ l^{-1}$
Nitrate: $1\ \mu mol = 1\ \mu eq\ l^{-1} = 0.014\ mg\ N\ l^{-1} = 0.062\ mg\ NO_3^-\ l^{-1}$
Sulphate: $1\ \mu mol = 2\ \mu eq\ l^{-1} = 0.032\ mg\ S\ l^{-1} = 0.096\ mg\ SO_4^{2-}\ l^{-1}$

Units for converting mass per unit volume to volume mixing ratios (at 20°C and 1013 mbar pressure)

SO_2:	1 μg (SO_2 as S) m^{-3}	= 2 μg SO_2 m^{-3} 0.75 ppb
NO_2:	1 μg (NO_2 as N) m^{-3}	= 2.1 μg NO_2 m^{-3} 1.72 ppb
O_3:	1 μg O_3 m^{-3}	= 0.5 ppb
NH_3:	1 μg (NH_3 as N) m^{-3}	= 1.21 μg NH_3 m^{-3} 1.72 ppb

Index